How to Build the Ultimate V-Twin Motorcycle Engine

by Timothy Remus

Published by:
Wolfgang Publications Inc.
PO Box 10
Scandia MN 55073

First published in 1996 by Wolfgang Publications Inc. PO Box 10, Scandia, MN 55073, USA

ISBN number: ISBN 0-9641358-4-1

Printed and bound in the USA

On the cover:

The beautiful blue and polished V-Twin is the property of Mark Shadley from Whitman, Massachusetts. Mark's V-Twin uses S&S flywheels to bump the displacement and Sportster rocker boxes to provide a unique look. The air cleaner is from Cycle Fab and the paint is by Auto-Tech.

How To Build

The Ultimate V-Twin Engine

Contents

Acknowledgements

It's that time again. Another book is finished and among the small but important jobs that always get left until the very end is writing the Acknowledgments.

This book has been an especially hard one to write because I never knew when to stop. Any topic in any chapter could have been a book in itself. And the deeper I got into the various topics the more I had to rely on my technical advisers.

The list of advisers might start with Dan Fitzmaurice from Zipper's. Dan is as knowledgeable as anyone building and racing V-Twins today, yet he gave freely of his time and expertise with no strings attached. Equally legendary in racing circles is Craig Walters of Walters Technology. Craig is quick with a joke, yet behind that "aw shucks" manner is an extremely competent engineer and computer wizard.

John Andrews is the man behind Andrews products, for 20 years a leader in the design of camshafts. A man who now sells his own program to help other camshaft manufacturers design the perfect cam lobe.

Lee Wickstrom from Kokesh MC outside Minneapolis, Minnesota, is the man who uses camshafts from Andrews and Accelerator programs from Craig Walters to assemble better V-Twins. Lee builds everything from restored Panheads to smokin' 96 cubic inch V-twins and they all run long and hard. From the West Coast I have to thank wild man John Reed, who by day designs parts for Custom Chrome and by night toils away on a motorcycle that really is the "Ultimate V-Twin."

I also need to thank Carl Morrow of Carl's Speed Shop and Jerry Wilhelmy from General Engineering for their interviews, and John Bryant from Motorcycle Engineering in Cleveland for a nice assembly sequence. For Shop Tours I am grateful to Randy Torgeson of Hyperformance and Rick Vandehaar from Axtell.

Though most of the photos in the book are my own, many come from ad agencies and the various companies listed in the Sources section. I must issue a collective thanks to all the people at all the ad agencies and media departments for finding the right material and sending it to me on remarkably short notice.

As I do more and more books of my own I come to rely on a very small crew for help when I really need it. That help includes transcribing interviews, which is the domain of Gail Fairchild. Jason Mitchell, another Kokesh employee, always helps me get those last minute photographs. For doing great layouts and working through long weekends without much sleep I rely on Mike Urseth. And for correcting my grammar and misuse of the English language I must of course thank my lovely and talented wife, Mary Lanz.

Introduction

How it all started

This whole Ultimate V-Twin business started when I realized how many people want to build their own motorcycle from scratch. The first Ultimate book - *How To Build The Ultimate V-Twin Motorcycle* - is intended to guide people in buying parts and assembling those parts into a complete running motorcycle. That first book has only one chapter dedicated to the engine, thus was born the idea of a book dedicated to engines.

What it is

You hold in your hands a book about buying and building a V-Twin engine. The focus here is on current Big Twin engines and all comments (unless noted otherwise) refer to evo-style Big Twins. It's not that there is anything wrong with those other and earlier engines, but rather the fact that it's hard enough to get a handle on all the parts available for the current Big Twin, much less the other engines.

There is an explosion going on in the V-Twin aftermarket. Companies both large and small are bringing new parts to the market daily. The new stuff runs the gamut from exotic new fuel injection systems to V-Twin engines with roughly the same displacement as the V-6 in my little Ranger pickup truck.

Though the exotic stuff has a certain sex appeal that's hard to ignore, I've tried to keep the book focused on "real world" parts. Most of the information is intended to help you buy or build a strong street engine of 80 to 100 cubic inches. The emphasis here is on street bikes with usable power and good reliability. Which is not to say we ignored the exotic engines. Chapter Eight is intended to answer your questions about blowers, turbos, NOS and engines with 179 cubic inches of displacement.

What it's not

This is not a book about competition engines, because competition engines *are* different (no matter what people might say). Drag strip engines run wide open for short periods of time. They don't need to idle through traffic or pull smoothly from 2000 rpm. And though assembly sequences are included in Chapter Nine, this is not an assembly manual. Which means you should go out and buy a service manual for your new engine. That way you will have the correct procedures, and the right torque and clearance specifications for your particular engine.

Like all books, this one is a compromise. If you distill it all down, this is an attempt to make readers think first and buy second. To help them avoid the worst pitfalls of over-eager builders of horsepower.

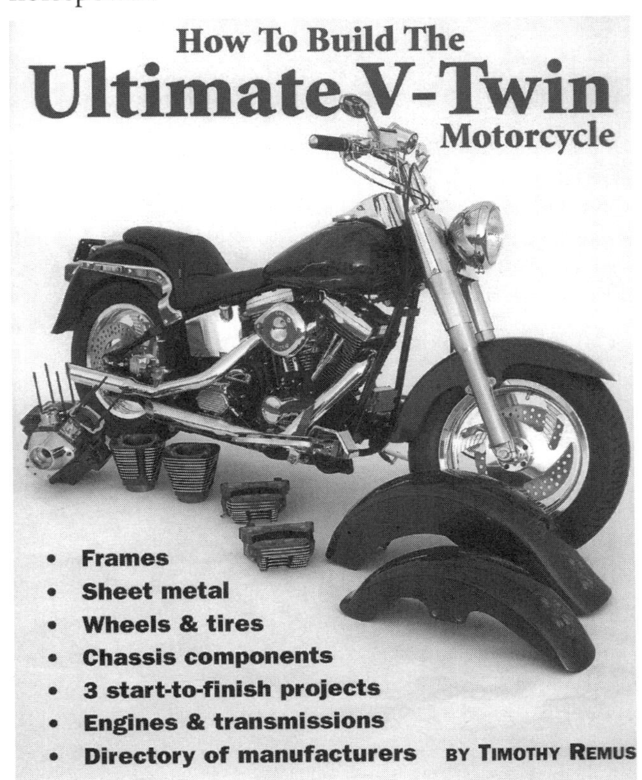

Chapter One

Design The V-Twin That's Right For You

How big, how fast, how expensive?

INTRODUCTION

It used to be easy. There just weren't all that many options. Most riders started with the V-Twin that came with their bike from the factory and started modifying. Sure, there were some aftermarket cases and cylinders, but most of those were intended for competition use. Street riders tended to stay within a limited set of parameters based on the factory engines from Harley-Davidson.

The problem today is deciding what to buy from

Designing an engine from scratch is no small task. You need to know how big, how fast (for how long), what color to paint it, how much chrome and billet to use and how much all that is going to cost.

among the vast array of possible components and complete engines. The other problem faced by the current rider looking to buy or build a complete motor is the fact that what was once considered exotic is now pretty standard stuff. 88 and 96 cubic inch motors are no longer a big deal. If you want real bragging rights you've got to belly up to the bar for the really cool stuff. Be sure to bring your checkbook because the really neat stuff is never cheap.

If you're building a cruiser bike on a budget and you want good reliability the Harley-Davidson crate motors represent a good value. If, on the other hand, nothing but the best will do, Keck and Hyperformance (among others) build billet engines that go well beyond 120 cubic inches.

A Kid In The Candy Store

With all the jeweled engines and components available it's hard to decide which parts to bring home. You have to begin by being brutally honest with yourself. How will you use the bike, where in the rpm range do you want the power, how fast can you afford to go and how long do you want this new motor to last?

The rider who wants to race from stop light to stop light will need a different engine than a rider who wants to cruise on the highway. As the size and horsepower go up the cost does too. Exotic engines and components cost the most and also cost more to assemble and maintain.

And don't overlook the visual considerations as you plan and budget for that new V-Twin. It might be tempting to buy a compete motor from Harley-Davidson or a company like S&S Cycle, but what about painting the engine? It's hard to do a nice paint job without disassembling the engine. And though the Harley motors usually come in wrinkle-black that won't do you much good if you want the cases candy apple red to match the rest of the bike.

Visual consideration brings up the subject of billet covers and accessories. Arlen Ness makes some truly lovely polished covers and air cleaners, but they aren't cheap and need to be considered as you work out the budget.

The "B" Word

Reality has a way of spoiling even the most modest of dreams. In this case reality is synonymous with the word "budget." You can't assume that the engine of your dreams can be had for five thousand dollars if that same engine will include paint, polish and plenty of billet goodies.

Life is often a series of trade-offs, and it's up to you to decide what you really need and what you can live without. First decide (based on riding style and your end use) what type of engine you need. If the engine that suits you can only be purchased as components, then be sure to include enough in the budget for labor. Be careful with the 'ballpark" estimate the mechanic gives you, the real cost always seems to be higher than the first rough estimation. Paint and polishing will add to that labor figure. Then add in the cost of necessary parts that might not come with your motor, things like the alternator, rocker boxes and ignition system.

As you total everything up the figure might surprise you. Now is the time to decide which you want worst, the polished billet outer primary or the trick single-fire ignition.

Entry Level

The least expensive engines are probably the crate motors from Harley-Davidson dealers and the nearly-

Most V-Twins look pretty much alike. Dave Perewitz managed a unique approach a few years ago with this dual carb Big Twin built with 2 front cylinder heads.

DISPLACEMENT CHART

STROKE	— BORE —					
	3-7/16	3-1/2	3-5/8	3-13/16	3-7/8	4
3-31/32	73.6	76.3	81.8	90.5	93.6	99.7
4-1/4	78.8	81.7	87.6	96.9	100.2	106.8
4-1/2	83.4	86.5	92.8	102.7	106.1	113.1
4-5/8	85.8	89	95.5	105.6	109.1	116.2
4-3/4	88.1	91.3	97.9	108.5	112	119.4
5	92.7	96.1	103.1	114.1	117.9	125.7
5-1/4	97.4	101	108.4	119.9	123.8	131.9

complete engines available from S&S. Also quite reasonable are the somewhat basic motors assembled by your local V-Twin mechanic from aftermarket parts.

Prices of the crate motors will vary with the dealership. Some mechanics can build an aftermarket engine from components for about the same money as the crate engine. Which is the right one for you depends on what you really want. As George Edward from St. Paul Harley-Davidson in St. Paul, Minnesota explains it," These Harley-Davidson engines are complete right down to the alternator, carburetor and ignition system, minus the module. Each one has been run on a test stand at the factory and must pass all the items on their check list before it's shipped." Also consider the fact that some insurance companies will insure the bike at a lower rate if it is powered by the complete Harley-Davidson engine.

Each Harley-Davidson dealership gets only so many motors each year on an allocation. If you're interested in a factory crate motor contact the dealer in your area. Motors are current-production only, which means you can't buy a complete Shovelhead from the dealer. Though new bikes are selling for list price, there is sometimes an oversupply of engines at the local level - so you might be able to get yours for less than list.

Engines from S&S or a motor built from components will almost certainly have more power than a Harley-Davidson engine. If you're interested in a factory crate

When it comes to visual considerations you're got to consider both sides of the engine. Note the billet coil bracket (which also makes a nice heat sink) and the polished and painted cylinders.

motor, get quotes from your local dealers (some will discount the list price) and then compare those to the aftermarket engines from S&S or the local shop.

More Power - More Money

More power generally costs more money, but the relationship between the two is not strictly linear. To quote Rob Carlson from Kokesh MC located outside Minneapolis, Minnesota, "You can get to a hundred horses fairly easily, but beyond that you have to start doing some major and financially challenging things to get the extra horsepower out of it."

So even if you do want extra horsepower understand that as you spend more and more there are diminishing returns. Exotica might be great but it's expensive and hard to have serviced if it breaks while you're out of town. Be sure to get the maximum value for the money you spend and understand that you never get something from nothing.

Some of the large displacement options like the Keck engine with its 4-1/4 by 4-1/4 inch internal dimensions make great horsepower and torque, but often won't fit in a "standard" frame. Many strokers and large cubic-inch engines are taller than a standard 80 cubic inch engine. If this is what you really need be sure it will fit in your intended frame.

Find A Shop

Unless you are ready to buy a complete motor or do *all* the assembly you're probably going to need help putting this motor together. In this "real world" of motorcycle construction you need to give as much consideration (or more) to the shop that does the bulk of the work as you do to the parts you buy. Find the shop early in the project so they can help you find the right combination of parts.

Pick the shop with care, someone with experience and a long track record. A shop or dealership that's been in business long enough that you can be sure they will be there if you need help one year down the line. Ask specific questions as to the cost for parts and labor and warranty. Some shops insist you bring the engine back to them for the initial firing, so they know all the oil hoses are routed correctly and that the engine isn't

The Feuling 4-valve heads use their own intake manifold. One that hangs a carburetor on either side for good performance and a more balanced appearance.

LEGAL MATTERS

To title a scratch-built bike in most states you need a MSO (Manufacturers Statement of Origin) with serial numbers for both the engine cases and the frame. When you buy cases or a complete engine from any legitimate aftermarket supplier you will get the MSO. Be sure it is filled out correctly and that any previous transfers (from one shop to another for example) are noted.

Because of the potential for legal hassles later a complete used engine might not be such a hot deal, unless you buy it from a reputable dealer or junk yard who can provide the necessary paperwork - all completed to the satisfaction of your state licensing authority.

For the same reason it's not a good idea to base your engine on a set of used Harley-Davidson cases. Without a MSO the state will not give you a title for the new bike. If you're putting a complete aftermarket engine in an already completed and licensed motorcycle you need to take the title for the bike and the MSO for the new engine to the state so the new engine can be noted on the title. You may also be required to bring the bike to the state Motor Vehicle Testing Station for an inspection (this is for Minnesota, your state might be different).

As long as we're talking about legal issues, you must keep the receipts for everything you buy for the new scratch-built bike and/or engine. In this way you can prove to the state that the parts are not stolen and also that the sales tax has already been paid. Sometimes people buy a set of new aftermarket engine cases and then a complete engine, minus cases, at the swap meet. Then they match the swap meet internals with the new, legal, external cases and you've got a cheap engine. Life is good until it comes time to prove to the state that you bought everything, including the pistons, flywheel assembly, cylinders and so on. Some of the state inspectors are smarter than you think. Even if the state doesn't give you a hard time it doesn't make good sense to support the people who might steal *your* new bike.

At the risk of sounding like your mother it's always best to be extra thorough in documenting everything you do. Keep perfect records and don't cheat on the paper work. A photo record of the project is a good idea too. That way you will be able to get a nice clean title - one that will stand up to the scrutiny of even the toughest cop on the darkest night.

MANUFACTURER'S STATEMENT OF ORIGIN
MOTORCYCLE CRANKCASE

The undersigned, S&S Cycle, Inc. (the "Corporation"), hereby certifies that the new motorcycle crankcase described below, the property of the corporation, has been transferred May 16, 1995
on Invoice No. 086307 to: KOKESH CYCLE
8302 HIGHWAY 65
SPRING LAKE PARK, MN
55432
PH.#612-786-9050

CRANKCASE DESCRIPTION:

Trade Name:	SUPER STOCK	Year Mfg:	1994
Model:	31-0001	No. of Cylinders:	2
I.D. Number:	U 0915X	Case Weight (Dry):	23

The Corporation further certifies that this was the first transfer of such new motorcycle crankcase ordinary trade and commerce.

S&S PROVEN PERFORMANCE
S&S Cycle, Inc.
Box 215 Route 2 Hwy G
Viola, Wisconsin 54664

By: (Signature) Samuel L. Scaletta
(Title) Corporate Officer

FIRST ASSIGNMENT

FOR VALUE RECEIVED, the undersigned hereby transfers this Manufacturer's Statement of Origin and the motorcycle crankcase described therein to:

When you buy a new motor or set of cases be sure you get a MSO, correctly filled out - you will need it to get title to the new bike.

allowed to build too much heat during the first few minutes of running.

Bore Or Stroke?

If you want to start a fight at the bar next time you're there just tell somebody that you get more torque from a big-bore engine than you do from a stroker. The bore versus stroke controversy is an old one and it isn't settled yet. Traditionally anyone building a V-Twin significantly bigger than 80 cubic inches would go to a longer stroke before considering a bigger bore. Though strokers are always popular, more and more shops are using big-bore cylinders of 3-5/8 inches or more to get the extra cubic inches (you will find more on this controversial topic in Chapter Two).

For now, consider that you need to determine more than just size and "power." You need to go one step further to decide the type of power you want and the best way to obtain that type of power delivery.

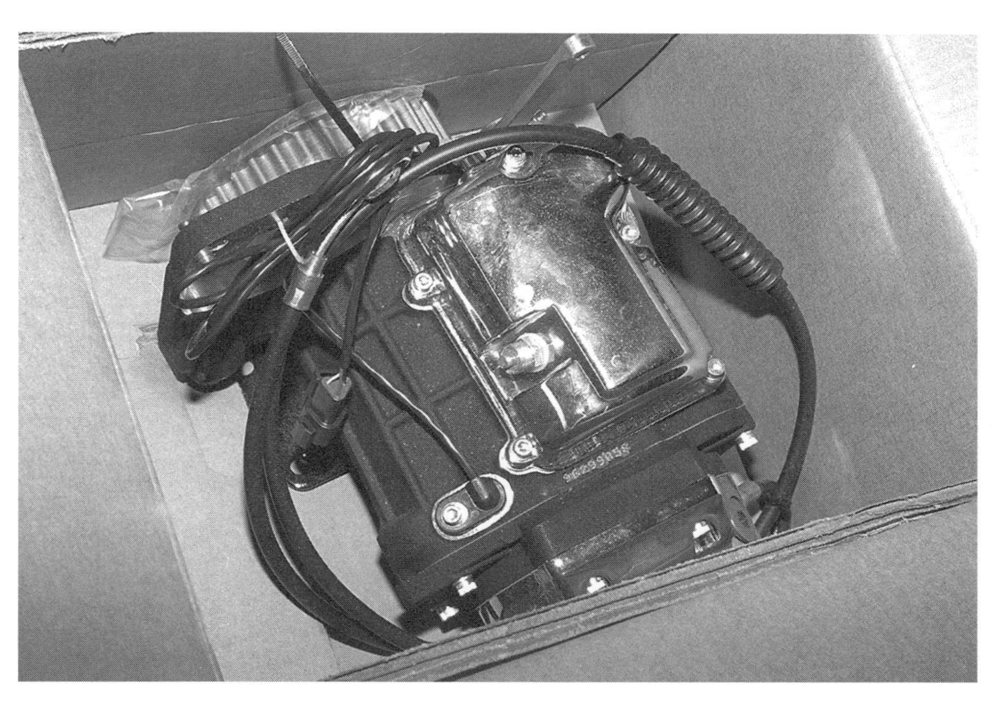

Harley-Davidson makes complete transmissions like that seen here. You can also buy bare cases and gearsets. In the aftermarket RevTech and Andrews make gear sets while Delkron and STD manufacture cases. The style of the case (soft-tail or rubber mount) is dictated by the frame design.

Interview, Dan Fitzmaurice From Zipper's

Dan Fitzmaurice from Zipper's is more than just a successful drag racer and merchant of speed equipment. Buzz Kanter, publisher of American Iron and Thunder Alley, calls Dan, "A cerebral racer, a guy who races as much with his mind as with his body. Dan is the kind of guy who always wants to know not just if it works, but why it works. He's got a natural curiosity and he's always trying new ideas."

Dan, what if I walked into Zippers and said, "I'm going to build a bike from scratch.." What parameters do you use to help figure out how much motor I need?

Before we determine the

The inner primary can be plain aluminum or chrome plated like this example from Arlen Ness, but it must be matched to your chassis/engine/transmission combination. Arlen Ness Inc.

cubic inches of the right motor for you we would define subjective data. We would decide what you are going to do with the bike. One client might have a Kenny Boyce frame and he's at the drag strip racing on weekends and he's a very aggressive rider. The next one that comes in might be a cruiser. It's going to be a really beautiful custom bike. He's not interested in going past 120 miles an hour, but wants to get from 0 to 80 in a hurry. We define the way the bike will be used first.

Then we look at what the weight of the motorcycle is going to be. And the weight of the rider. Is it going to be ridden two-up or will the owner always ride solo. Now we've got a picture of total tonnage and usage. That in turn determines how the motor is going to be cammed, exhausted, and the type of cylinder heads and/or probably the compression ratio.

We're trying to help the customer define what his expectations are, and if he doesn't know, we're going to ask enough questions that we're going to draw the picture for him.

What if I explain to you that I'm not going to drag race but I want to have a lot of snort when I come out of the hole. I'm going to ride single. I'm defining the somewhat typical rider. I probably won't ride 120 miles an hour ever. I won't do much real sustained high-speed operation. The longest trip will probably be 500 to 1000 miles one way. But I do a lot of running around. bar to bar, so to speak. With those parameters, help me define how many cubic inches I need and where we go from there.

What we'd probably do is set you up with a larger than 80 cubic inch displacement engine. You'd be buying everything new anyway so the price isn't a factor. That'll give you the torque you're looking for, that certain feel. And also, if you build a $20,000 to $30,000 motorcycle, depending on what you're doing, you might have a problem socially with your friends saying it's only 80 inches. At least that's something we hear a lot. This way you've got a 97 inch motor or a 102 inch engine. We do it with a combination that keeps the stroke down. We typically don't run much more than 4-5/8 inch stroke. Based on the criteria you're mentioning we're probably talking 4-1/4 to 4-1/2 inch stroke. Piston speed is minimal that way - for good longevity. This would allow you to have a nice engine that's reliable. And you have the needed items on your bike that makes it unique.

So you would rather get the extra cubes from a larger bore versus a longer stroke?

It is our preference to get the extra cubes from the bore, because that way we get reduced piston speed and longer piston skirt length - which in turn gives us more stability, a more rigid cylinder. I would comment that although we don't do a lot of them the S&S 96 inch motor is a really good value. For the dollar you can't replace that with any other engine of similar size. It typically runs a 4-5/8 stroke and 3-5/8 bore. It's not the fastest thing out there for its size, but it's a good value with a proven track record.

Let's talk about cases. Do you have a preference as to cases?

I think everything out there is pretty good. What you're looking for, for the guy building his own engine as well as an engine shop, is how much work is involved in those crank cases to make them finished.

We try to buy cases that don't need much in the way of finish work.

If we buy something that requires a lot of hand work it adds to the cost for the customer. Whereas, if you can buy something where a lot of the work is finished, it allows you to focus on inspection and fit and making sure everything looks good.

You prefer cast iron cylinders?

Yes, for rigidity and to keep the cylinders round for a good ring seal.

You don't have a problem with the different expansion

Ed Kerr chose paint and polish for the V-Twin in his custom Softail. Note the unique one-off aluminum air cleaner made from sheet aluminum.

rates of the cast iron and the aluminum pistons?

No. You may want to mention that there are several different types of cylinders. You've got ductile-iron, racing stuff, which we don't recommend on the street because it does not transfer heat well. The cast iron is much more conductive and better at displacing the heat. Cast iron engines have been built for years. There's not much reason to run the ductile cylinder on a road engine.

Do you typically install forged or cast pistons? What about the whole issue of fit, can the cast pistons be fit tighter than forged?

At one time you had to buy a cast piston to get one with a high silicone content and good expansion rates. But now there are quite a number of low-expansion forged pistons available with very high silicon content. On street engines we can fit the pistons nice and tight provided that the rest of the combination is correct. You need a good carburetor with the proper fuel curve, no lumps or lean spots.

You said that when you use cast iron cylinders, the break-in is real important, the heat cycles, can you expand on that?

The engine will make a tremendous amount of heat when it first fires. So much heat that damage to the engine may result if it was fired the first time and left running (people also need to be sure the oil pump is primed before firing the engine). This is true whether the cylinders are aluminum or cast iron, but especially true with the cast iron. In discussing this topic with other engineers we have found that it is the scrubbing effect of the new rings against the finish on the cylinder wall - it's incredible how much heat it can produce in 30 seconds. It would be very good to highlight that any time any engine kit has been assembled, even if it's using stock aluminum cylinders, you should be very attentive to the heat that's created the first few times the engine is fired. I recommend that people start the engine, let it run for just a very short period, and then let it cool down before starting it again.

How many times do you go through these cycles, what's your procedure? How quickly do the running periods get longer?

We do this about six times. The first cycle might only last thirty seconds or less. The second cycle would be just under a minute and the third cycle would be maybe a minute and a half. Then you start to see the normal warm-up come into place. This is a very important part of the early break-in. There should be good ring seal. You won't scuff any parts this way. There's not too much heat too quick. It's probably overprotective, but you just spent all this money so why not?

How about flywheel assemblies? I suppose there are only a few fly-wheel assemblies out on the market.

Anytime we're doing a scratch-built engine I would recommend that the flywheels come from the aftermarket, like S&S. They offer good shaft selection and they are very high quality parts. They have done a great job with that end of the business. It's where they started a long time ago. Truett & Osborn now have some fly wheels that are made on CNC equipment (computerized mills and lathes) and it looks like pretty good stuff. They don't have the selection available that you would through S&S. For shaft availability S&S is pretty hard to beat.

So now we basically have flywheels and cases and cylinders with pistons. When I choose my heads, how do I pick the best ones for my project?

The heads should have been bought early in the project, because in order to build the crank shaft you need to know what pistons will be used. And in order to buy the pistons you need to know what the combustion chamber shape is going to be and the valve sizes. We're essentially building from the top down, not the

In the past few years more and more "extra large" V-Twins have come on the market, including this 120 cubic inch billet engine from Keck Engineering.

bottom up.

Please define effective mean compression? And why you prefer this figure to the simple static compression figure often used?

This is the final compression ratio after considering the cam timing. For example, an 80 inch engine with 10.3 to 1 static compression and a Red Shift 575 camshaft has 8.8 E.M.C. which is good. Switch to an Andrews EV3 camshaft and the E.M.C. changes to 9.6 to 1, which is too much for a road engine. If you start with lower static compression the EV 3 is a great fit - so it's a matter of matching the parts.

You said you like to see people do a mock up of the engine, put it all together and rotate it?

Yes. Make sure that everything that reciprocates and rotates in the engine has enough clearance and fits properly. Check your static compression ratio. Make sure your valves aren't too close to the pistons and also check the squish area in the heads.

This means mock it up just like it's assembled to run, with clay on the valve cut outs on the pistons. Except the valve springs, you want to use a light spring so you don't have deflection of rocker arms and pushrods. All these things assure you don't have to go back in and do things twice. For home builders this is a must. I tell them to have patience and go through the mock up stage until everything fits before the final assembly.

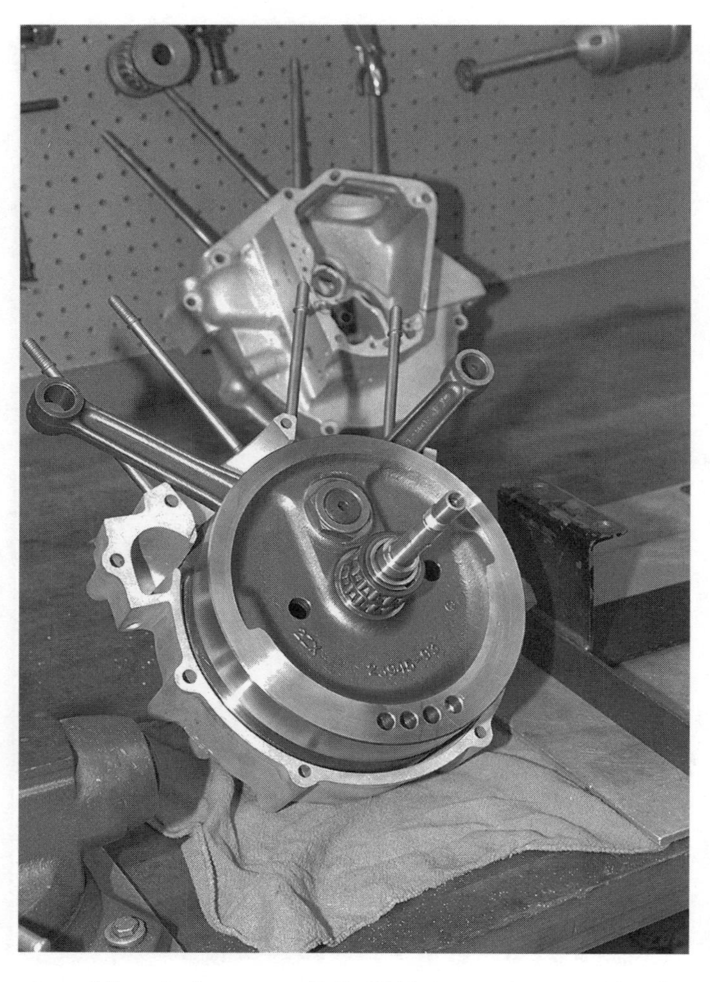

Assembling the bottom end of a V-Twin requires specialized tools and training. Even for "hands on" riders this is one part of the build up that's better left to professionals.

What about static versus dynamic balancing?

I feel that in a big twin engine, based on the criteria we talked about, if we have a conscientious technician balancing our crank shaft, statically or dynamically, we're going to get a good job. It's going to fit the bill. I'm more interested in conscientious people doing the work than one versus the other. If we turn the thing to 9,000 rpm then maybe we'd consider balancing dynamically first. On the street it's just not an issue.

You made a comment about torque versus horsepower. Could you expand on that and talk about components that would give us those characteristics?

Based on your engine criteria, torque, from our perspective, is the most important thing. That's what you're going to feel and taste every time you ride this bike. What we want is a broad, smooth torque curve. So that whatever gear you're in, however much weight you're packing that day, whatever the case may be, you have good clean throttle response that's usable. Your engine will like it. Your drive line will like it. Machinery talks to you, whether you're grinding a part on a precision machine or pulling away from the stoplight on your Harley. With a good engine you can hear the exhaust note and know it's right.

So I want a lot of torque and I want it relatively low in the rpm range, in a range where I'm riding and spending a lot of time?

In the original discussion with the customer we try to decide what we're going to build and where the power range will be - so it's not a one-size-fits-all situation. But again, we go back to the smooth torque curve. That gives you good power over a very wide rpm range so if you vary your riding style the power is there.

So how do we chose components that give us those characteristics? How do we avoid buying the wrong stuff?

It's a combination. First you need to determine the camshaft, cylinder head and combustion chamber size, which in turn gives us static and effective mean compression. Then figure out the carburetion and exhaust pipe selection. The keys to the whole thing are: induction and the exhaust design, the camshaft and the combustion chamber shape and efficiency. Really what you're talking about when you talk about a "kit" is those items. That selection of all of those things, each change in it, directly affects the others.

So there is no magic bullet?

No magic bullet. Unless we build everything the same, but in this market that doesn't seem to happen. A Harley-style engine is probably built in more different variations and sizes than any engine in the marketplace. Part of the reason for that is how easy it is to change strokes and bores. The Y-style intake manifold presents challenges that are not experienced with other motorcycle or automotive engines.

So from your perspective it makes sense to buy components, at least carburetor-head-camshaft components, as kits. Because if a reputable manufacturer has put together the kit, it's a combination that will work together and that's the most important thing. You still need to be sure it matches your expectations, in terms of the torque curve for example.

What you don't want to buy for this engine you're trying to build, is raw parts. If you do you're buying a development job. If you buy combinations that are unproved you're going to have a lot of development time in sorting everything out and probably a very high cost. That's something to think about when you want to do one of these new, revolutionary big inch street bike engines.

It might be repetitious, but I'd like to touch one more time on some of those horsepower/torque graphs we talked about and the folly of looking for the big number at the dyno shoot out?

Probably the thing that shows up the most in some of those big-number engines with small output is what we refer to as negative torque. This is a situation where the power comes up on initial throttle response and then actually dies. Then 1500 rpm later it may recover and start accelerating again. The rider of that particular engine might refer to it as a "pipey" or "cammy" engine where it could be a poorly-designed exhaust pipe, that may look beautiful but is really creating a negative pressure that backs up into the combustion chamber. It could be camshaft selection. It could be low static compression (or effective mean compression), or carburetion problems. There are a variety of forms it comes under.

The emphasis should be focused on making good power and torque over a wide rpm range. Torque, whether you're a racer or a cruiser, gets you there first.

So it's wrong to look at the peak horsepower numbers at 6,500 rpms.

Of course we want good numbers there, but if that's our benchmark for determining if we're doing a good job or not, that's a bad deal.

In terms of mistakes people make when assembling motors, you say they're not careful enough about dirt?

I wouldn't refer to it as dirt as much as "foreign contamination" which can come in a variety of forms. It's disguised as paint, primer, rust, material left from the chroming process, powder coating, glass beading. It comes through engine covers that are chrome, oil tanks, gas tanks, crank cases. It's a disaster when it's left inside the new engine and it happens way too often.

Any final comments you'd like to make. Mistakes people make or things they could do better.

People should spend more time in the selection process of what you're going to do. If they can engine-model on a computer first, that's great (*Editor's note: Dan means use something like the Accelerator program mentioned later in the book to examine how all the parts work together*). If not, then they should contact someone who can help and give advice as to which parts to buy and use. More time spent on the front end with some planning may same them thousands of dollars. And they need to avoid powder coating parts that are already finished. The heat changes things, like the size of the cylinders and the fit of parts that were already machined.

I would say new builders should be careful to make sure everything is very clean, that they have good preparation. The key to quality engine building is good concentricity and perpendicularity. They must do a good mock-up. Finally, good attention to detail during the assembly will ensure success at the end of the job - and save lots of money in the long run.

A long time drag racer, Dan Fitzmaurice from Zipper's understands high performance V-Twins, both on the track and on the street.

Chapter Two

Cases, Cylinders, Pistons And Flywheels

The heart of the matter

INTRODUCTION

Before ordering cases and cylinders you've got to decide exactly how many cubic inches this motor will be, and how those cubes will be obtained. "Stock" 80 cubic inch engines have a stroke of 4-1/4 inches and a bore of 3-1/2 inches. They say you can't beat cubic inches, but they don't tell you the best way to get those extra cubic inches.

You can have a 96 cubic inch engine with a stroker 4-5/8 inch flywheel mated to a 3-5/8 inch bore. Or

Though some cases are cut from billet aluminum, most are cast - like these at S&S - from 356 aluminum alloy heat treated to T6.

figuration best meets your needs.

Before beginning work on an engine most professional engine builders tell their customers to consider what kind of riding they really do, and where in the rpm range they want the power. Use those answers to help determine the number of cubic inches the engine should be and which configuration of bore and stroke best fit your unique situation.

You need to decide how big and how powerful an engine you intend to build and kind of work backward

Cylinders can be made from cast aluminum with cast iron liners, cast iron and even ductile iron for (mostly) racing applications.

you can have a 97 cubic inch engine with a stock, 4-1/4 inch stroke and a 3-13/16 inch bore. They aren't the same, which isn't to say one is bad and the other is good. They're both good, just different. With different prices, different power delivery, and potentially different longevity.

Though "strokers" have always been popular with V-Twin riders there is a new emphasis on big-bore cylinders as a means of gaining cubic inches. Proponents of this school often cite the fact that a shorter stroke engine tends to have better reliability - at least in the sense that piston speeds are lower at a given rpm, so ring and piston life are often improved. Short stroke engines are commonly thought of as being quicker to rev than a similar engine with a smaller bore and longer stroke.

The bore versus stroke controversy is another one that will always start an argument. "Strokers have more torque" someone will say, "because the longer stroke means the piston is pushing on a longer lever." "But big-bore pistons present a larger surface for the expanding gasses to work on," someone else might add. There is no absolute right or wrong answer. You can use bore, stroke, or a moderate increase in both to get those extra cubic inches. Ultimately you have to use your own experience, combined with the input of people you trust, to decide which bore and stoke con-

If you're spending lots of money to make it fast, spend a little extra and make it look good too. This V-Twin at Minneapolis Custom Cycles is based on Delkron cases painted red, with polished and painted cylinders, and heads that are blue.

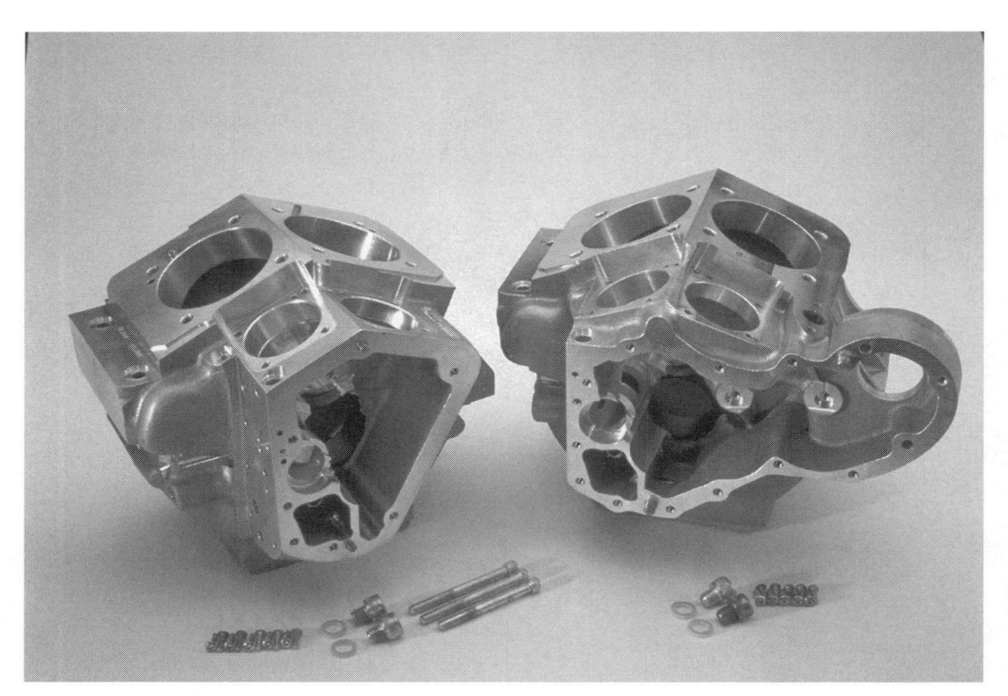

Manufactured with permanent molds, these Delkron cases feature heavy walls, reinforcement of any weak areas and an inspection plate on the bottom.

from there. Remember too that if you stay with conventional dimensions and parts you save yourself time, trouble and money.

Nearly all cases, cylinders and heads, up to a bore size of 3-13/16, will use the same cylinder-stud bolt pattern. Most of these parts are compatible with most others. As you move up to bore sizes of four inches and more, however, a number of different bolt patterns are used. Most shops still use through bolts in the larger bore sizes while others bolt the cylinders to the case and the heads to the cylinders.

If you want an engine with a larger bore size it will not only cost more, but you will also have to be sure the cases, cylinders and heads are all designed to work together. Many of these big-bore brutes have the lifter bores offset farther to the right to make room for the larger cylinders. This adds one more variable because now the pinion shaft needs to be longer than stock (longer pinion shafts are commonly available).

None of this is intended to discourage you from building, or having built, a motor that's significantly different from all the other V-Twins out there. It is intended to make you think about what you really want - and to make you understand *all* the costs and trade-offs involved in designing and building an engine (for more on really huge V-Twins see Chapter Eight).

Before you go out and begin ordering parts remember the comment of Dan Fitzmaurice from Zipper's: "The heads should have been bought early in the project, because in order to build the crank shaft you need to know what pistons will be used. And in order to buy the pistons you need to know what the combustion chamber shape is going to be, and the valve sizes. We're essentially building from the top down, not the bottom up."

Engine Cases

The engine cases you use for your personal Ultimate V-Twin will form the foundation for the rest of the engine. They not only support the flywheels and encase the oil, but are the attachment point for the cylinder studs and ultimately the heads.

There are no new Harley-Davidson cases available

Merch cases are cast from 356 T6 aluminum and can be purchased as bare components or as part of a kit with cylinders and heads. Bore sizes to 4-1/4 inches can be accommodated on some models. Mid-USA.

(unless you have an old set to turn in) so nearly any cases you start out with will come from the aftermarket. Most of the aftermarket cases have addressed and corrected weak areas in the factory cases. As Scott Sjovall from S&S said during our interview farther along in this chapter, "There are a lot of good strong cases out there."

Most cases are made from cast aluminum which is then machined on sophisticated CNC equipment. The aluminum is commonly 356 T6, which identifies the alloy and the heat treating. The upper end of the case market is occupied by the beautiful billet aluminum cases, which can often be mated to billet cylinders and heads.

If you intend to build an 80 cubic inch motor then there are a number of engine cases to chose from. Most standard aftermarket cases will accept a stroke of 4-5/8 inches (some up to 5 inches) and a cylinder with a bore up to 3-13/16 inches

Unless you're an experienced V-Twin mechanic you will probably use a shop or trained mechanic to assemble the bottom end. In which case they will likely recommend the brand of cases they prefer.

Note: When I asked the technicians at S&S if they ever had trouble with the threads on their cases they said some of those threads are "formed" rather than cut. This forming is almost like creating an extruded thread, and results in threads which are stronger than the more common threads cut with a tap. (If you look closely you will see that instead of a single ridge between the valleys, the formed threads have a "double ridge" between the valleys.) Sometimes the male threads will fit somewhat snugly in the formed threads. But the real trouble comes when people either chase the treads with a tap, in the belief that there's something wrong, or they use a chrome plated bolt or stud which can gall or lock in the formed threads because the chrome makes the bolt slightly bigger.

When you buy the cases, be sure they're well matched to your intended use. If you want big jugs, inquire of the case manufacturer regarding the maximum bore size the cases will accept and any special-

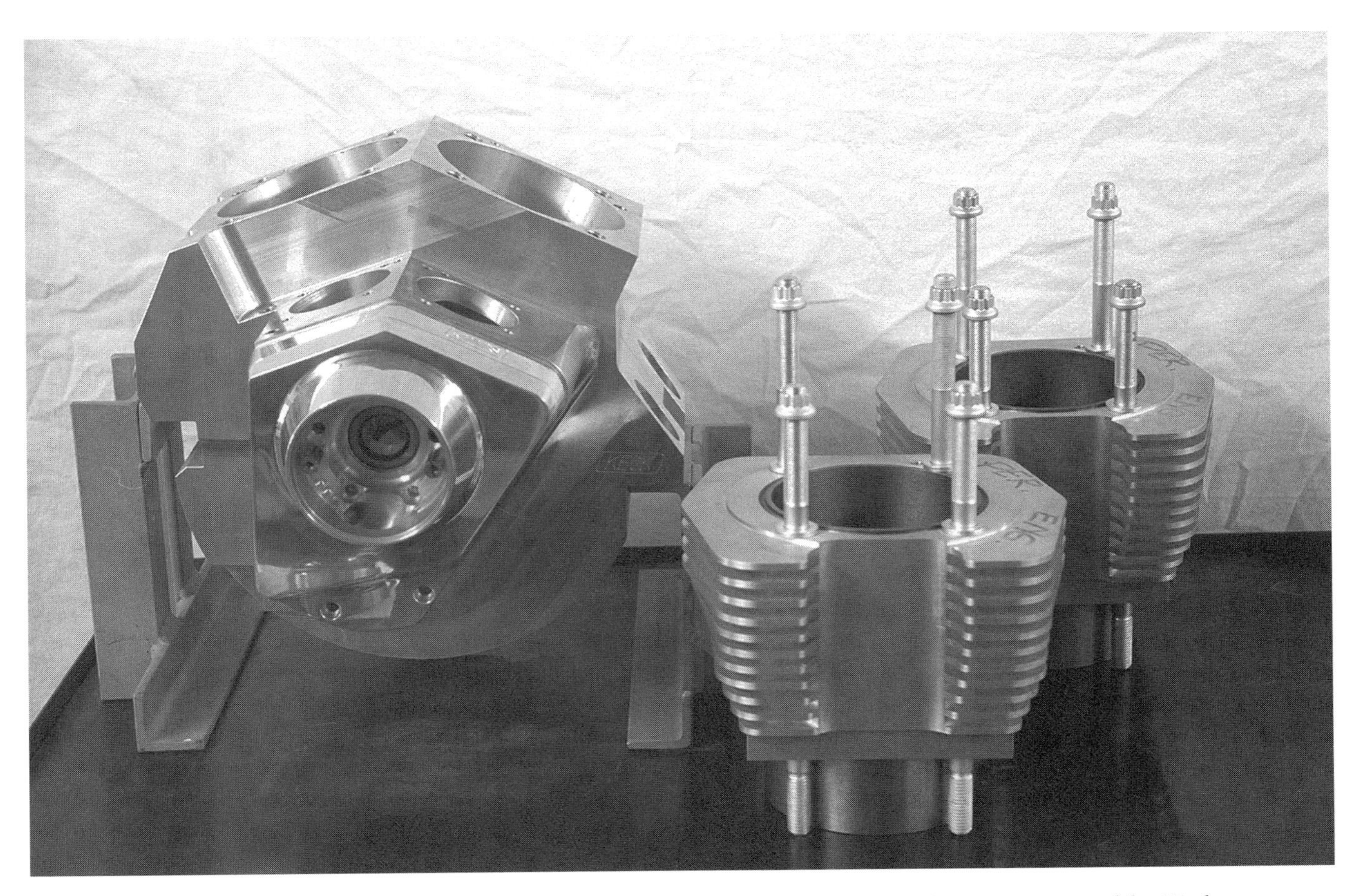

Billet cases and matching cylinders with their own unique look are available from companies like Keck.

S&S is one aftermarket company who manufactures engines that are very nearly complete (rocker boxes and cam covers are now available from S&S).

order considerations.

Some case manufacturers will bore the cases for cylinder spigots bigger than the cases were originally designed for, before they leave the plant. S&S, Merch Performance Incorporated and some others offer a matching set of big-bore cylinders in kit form designed to work with their cases.

Many manufacturers have big-bore cases available that will accept bores of 4 inches and more. These cases usually use a different stud pattern, which affects your choice of heads as well. Cases designed for a bore of 4 inches and more often have the lifter bores offset to the right (to make room for the big cylinders) which means you must use a longer pinion shaft.

Note: 1984 and up, evo-style cases are typically divided into one of two different versions. The 1984 to 1991 model takes a different oil pump and vents the crankcases differently than do the 1992 and later cases. S&S notes that there are more subtle differences as well, so the builder should be sure to match the "year" of the cases to the oil pump.

Most of the cases out there are "good" cases, and will stand up to all but the most insane street use. People do complain however, that a few brands require extra work to make them truly usable. This extra work might be as simple as cleaning up some threads or as complex as machining the deck surface so it is level on both sides.

Ask around at the various shops and the manufacturers to find out which set of cases will provide the strength, internal dimensions, frame-fit and price you are looking for.

Most STD cases are cast from 356 aluminum though special order cases can be cast in the stronger B358. Among the many cases they manufacture are these generator style cases for Evo barrels and heads.

The cases you buy need to be part of the overall design for the motor. They must work with the cylinders which must in turn be matched to the heads and pistons. Each part needs to fit all the other parts, your budget, and your intended use.

Engine Case Buyers' Guide

Delkron Cases

Delkron makes a number of different cases, everything from cast, late model Evo-style cases to billet four-cam exotica. Most riders and builders will be interested in the cast, Evo-style cases which accept a bore size up to 3-13/16. Bill from Delkron explained that even though the cases are cast from 356 aluminum they are cast in permanent molds as opposed to the more common sand-cast molds. According to Delkron, this means the aluminum is poured at a lower temperature which results in a more dense molecular matrix and a stronger end product.

If you're the kind of rider who needs maximum cubic inches, Delkron makes another set of cast Evo style cases with offset lifter bores capable of utilizing a 4-1/4 inch bore. And if you need still more, then there's the four-cam billet cases with raised deck and a maximum bore size of 4-1/2 inches.

Merch Performance

A Canadian company, Merch has long manufactured a variety of cast aluminum V-Twin components. Only in the last few years however, have their parts been readily available in the lower-48 through companies like Mid-USA and a few others. The Merch cases are cast from 356 aluminum heat treated to a T6 specification. Standard cases will accept a bore up to 3-13/16 inches, though 4 and 4-1/4 inch bores can be accommodated as well. The 4-1/4 inch bore cases are designed with offset lifter bores to make room for the larger cylinders. As a result, these cases take a longer pinion shaft. Cases can be ordered bare (with right race installed), or as a short or long block mated to Merch cylinders in various dimensions. If you want to build a '92 and later style Evo, then order your cases ready to accept the late model oil pump. Merch cases are designed to fit all factory frames, feature rolled threads for additional strength, and will accept a stroke up to five inches.

STD

STD Development Company has been making cases for V-Twins since 1975, and in all that time manager Doug Linscott has seen only two broken cases. "Our cases are sand cast," explains Doug. "Some people say billet is stronger, but ours test to 46000 PSI tensile strength. We feel they're the strongest cases in the industry. The strength comes from the material, ours is first-run 356 aluminum, and the heat treating that happens after the parts are cast."

Though you can buy a "standard" STD case, Doug explained that STD is a custom manufacturer. "People call me up and tell me how big they think they want the motor and we kind of go from there. We can tailor the cases to a specific project, including bore sizes all the way up to four inches with a half-inch raised deck." STD cases will accept a variety of cylinders, including large-bore cylinders from either Axtell or Hyperformance (both these companies can be found in the Sources).

STD cases are designed to fit a stock factory frame without modification - and without being a clone of the factory cases. They recommend flywheel assemblies from S&S for most applications, and occasionally

Seen here waiting for final machining, S&S flywheels are manufactured from forged steel and come in various diameters and a large range of strokes.

You have to pick the heads early in the project because they affect your choice of pistons which in turn makes a difference to the person who balances the flywheel assembly.

Truett & Osborn Company. Because STD is essentially a custom manufacturer they are happy to discuss your project before you get started. "Unless you already have everything figured out, give us a call so we can talk over what's best for your application," explains Doug. "We can help you design a short stroke engine that revs real quick or a long stroke, tractor motor. People shouldn't figure they know everything. I tell them to 'drop a dime' and give us a call, they might learn something if they do."

S&S Cycle

Perhaps the best known of all the aftermarket engine-component manufacturers, S&S offers their "Super Stock" cases for sale in permanent mold 356 aluminum, heat treated to a T6 specification. Standard cases accept either the stock factory cylinders or the 3-5/8 inch S&S big-bore cylinders. The new S&S "race" cases will soon be out that utilize offset lifter bores and a longer pinion shaft to accommodate an even larger bore. By the time you read this they will have both their new generator cases as well as the new Sportster cases, that accept a bore up to 4-3/8 inches, ready for delivery.

S&S cases fit all stock frames (are also available to fit early Shovel-type frames) and typically ship with the Timken bearings, studs and the oil pressure sender. Standard cases can accept a stroke all the way to five inches and are clearanced for the heavy duty S&S rods. These cases also feature an oil drain in the bottom of the cases for complete oil changes.

Sputhe Engineering

Sputhe Engineering makes cast Nitralloy cases in a num-

Pistons too come in a vast variety of styles, shapes and dimensions. The shape of the dome must work in concert with the shape of the combustion chamber. Some feature a pin location that's very high in the piston so that a longer connecting rod can be utilized.

ber of variations, from stock replacement cases to big-bore and/or raised deck examples designed for maximum cubic inches. Sputhe cases will accept a bore of up to four inches, though a bore of 3.78 inches (the size of Sputhe big-bore cylinders) is often combined with a stroke of 4.25 inches to create the Sputhe 95 cubic inch V-Twin.

Flywheels

The flywheel assembly you chose for that new V-Twin will likely come from one of three sources: Harley-Davidson, S&S or Truett & Osborn. If your engine comes as a partial kit then the flywheel assembly will come with the cases and other parts, and may even be installed. Most case manufacturers will recommend a particular flywheel assembly.

The S&S catalog lists a rather large variety of flywheel assemblies, including wheels of two diameters, 8-1/2 or 8-1/4 inches. While most stock V-Twins run a flywheel of 8-1/2 inches the smaller wheels are useful for stroker engines. Stroker engines bring the piston much closer to the flywheels at the bottom of each stroke. Because of this it is often necessary to use a piston with a short skirt area, which in turn makes for a piston that wobbles in the cylinder. The smaller diameter flywheels (which S&S recommends for 98 and 103 inch Sidewinders) means you can build that stroker motor and still use a more stable piston with a longer skirt.

Truett & Osborn

Truett & Osborn in Wichita, Kansas offers flywheels from 4-1/4 inches all the way up to a full 5-1/4 inches. These flywheels are made from ductile iron and heat treated before being machined. Complete flywheel assemblies are offered as well, including special order items with longer pinion shafts.

Harley-Davidson

Harley-Davidson offers their stock, 4-1/4 inch flywheel assemblies for sale from any dealership. George Edwards of St. Paul Harley-Davidson feels these are very high quality parts, "The factory has invested in plenty of new tooling in the last few years. The flywheels for example are a three piece assembly where

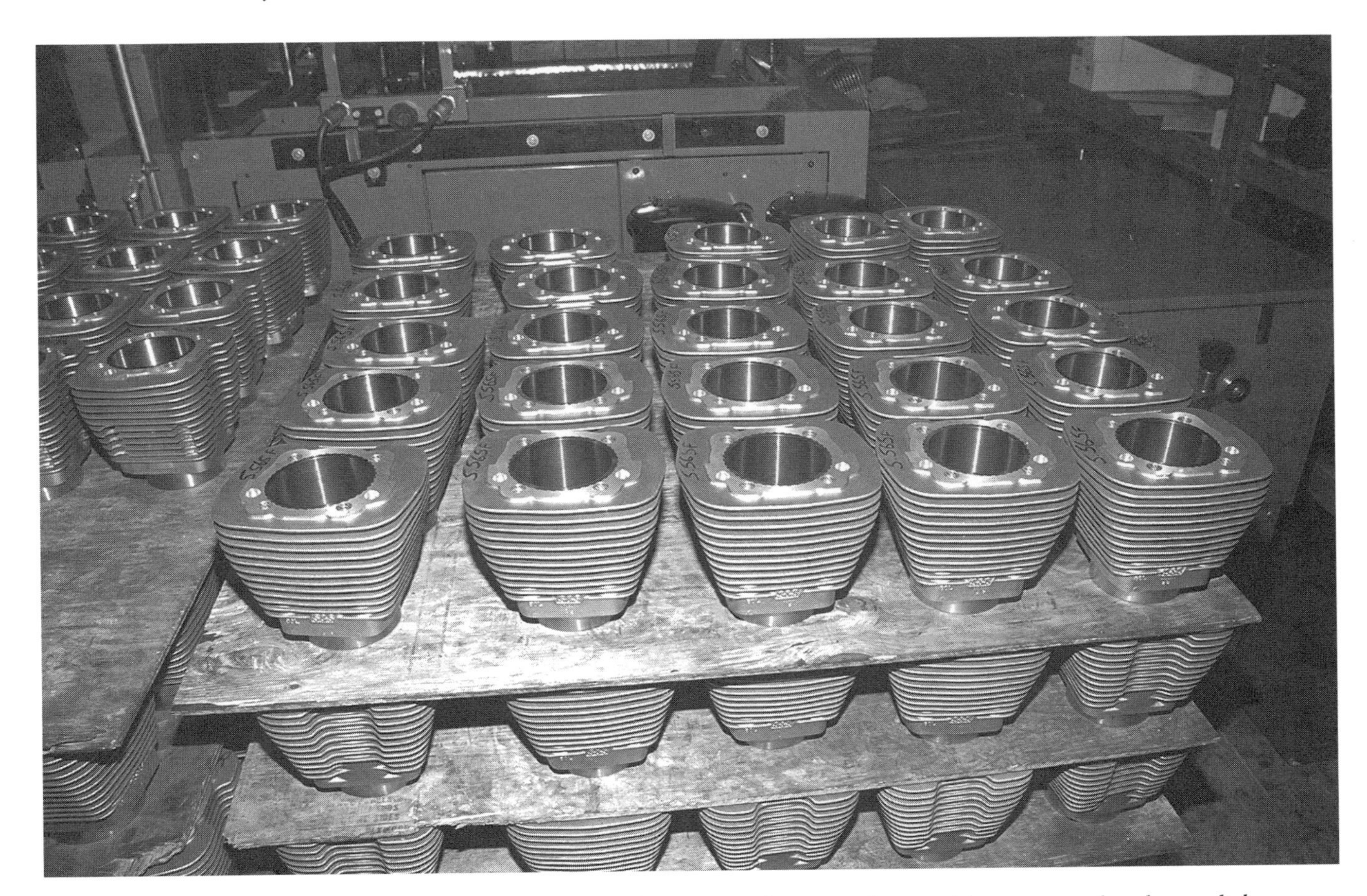

These cast aluminum cylinders from S&S come with cast iron liners and are said to run cooler than solid cast iron cylinders.

the aftermarket assemblies are five pieces with separate pinion and sprocket shafts that bolt into the flywheel on either side."

Cylinders

For street use, cylinders break down into two basic materials: aluminum with a liner made from cast iron, or solid cast iron, (in a few instances ductile iron is used on the street). As always, each material has advantages and disadvantages and everyone seems to have an opinion.

Advocates of cast iron cite the fact that cast iron doesn't "grow" with heat as fast as aluminum so the compression ratio cold is the same as it is hot. Cast iron cylinders are said to be more stable and distort less as well, meaning a better ring seal and improved oil control. The down side to all this cast iron business is its inability to give up heat to the surrounding air. How big a problem that is depends on who you talk to.

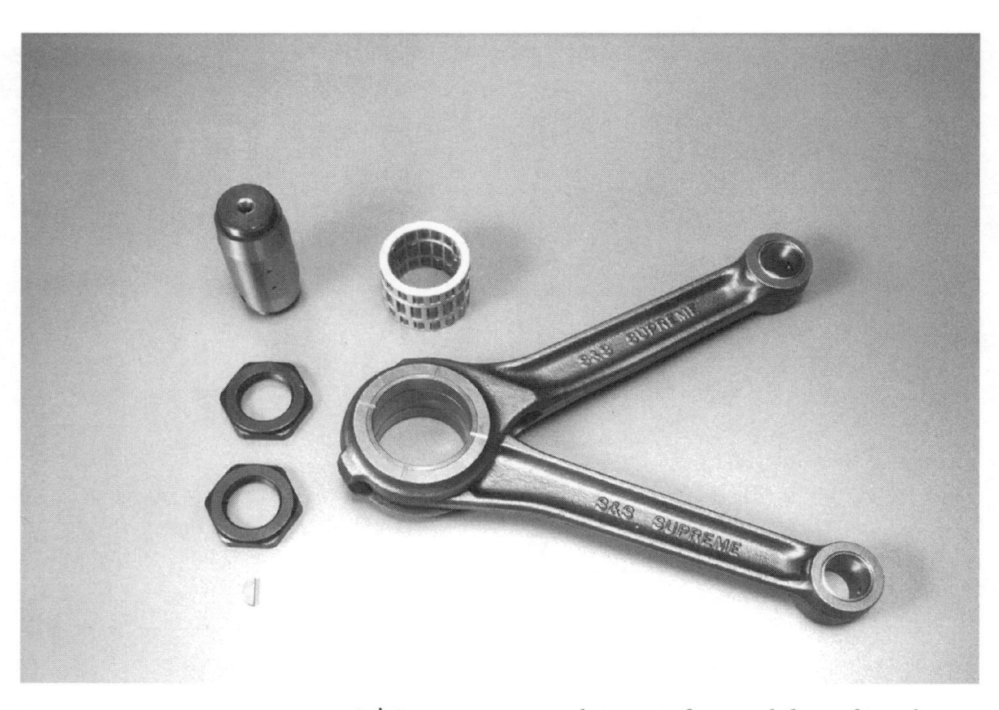

S&S connecting rods, manufactured from forged 4140 chrome moly steel, come in Heavy Duty and Supreme designations. While the heavy duty rods are fine for most street applications the Supreme rods are meant for racing engines.

Ductile iron cylinders, with tensile strength up to 90,000 psi, are cut from billets of iron like these at Hyperformance.

Designers in the aluminum camp like the non-ferrous material for its light weight and willingness to conduct heat to the air, which they feel results in lower combustion chamber and oil film temperatures, and prolonged engine life.

Dan Haak, an engineer at S&S, explains that they take the growth of aluminum into account when they design the parts, so it presents less a problem than people might think. He feels that because the expansion rate of the cylinder and piston are nearly the same, the piston fit cold can be tighter than it might be with a cast iron cylinder and aluminum piston. This means less noise and an improved seal, cold or hot. He also feels that the lower cylinder temperatures seen with aluminum cylinders have a significant impact on reducing the tendency to detonation and piston scuffing.

Cylinder Buyers' Guide

Axtell

Axtell cylinders come in two materials, cast iron or ductile iron (see Shop Tour). Most of the cast iron cylinders go out the door in sizes up to 3-13/16 inches, though Axtell can bore the cast iron up to maximum of 4 inches. This requires a different bolt pattern than stock. Ductile iron cylinders intended for competition use can be bored all the way to 4-1/2 inches. Cylinders from Axtell can be shipped with or without pistons, which are available in a wide variety of styles.

H-D

If the engine you're building is an 80 cubic inch mill consider stock H-D cylinders and matching pistons. The cylinders are cast aluminum with cast iron sleeves, and commonly go well past 50,000 miles without any trouble.

Hyperformance

Hyperformance manufactures ductile iron cylinders with fins that will accept a bore size up to 5.1 inches. In addition they handle billet and cast aluminum cylinders from other manufacturers which are also intended for larger-than-average bore sizes.

Merch

A variety of big-bore cylinders in cast aluminum with cast iron liners are available from Merch. Bore sizes start at 3-13/16 inches, and go all the way to 4-1/4 inches for a square 120 cubic inch V-Twin. Cylinders can be ordered bare or with matching JE forged pistons.

Sputhe

Sputhe Engineering has manufactured cast aluminum cylinders with cast-in sleeves since 1977. Cylinders are cast in 383 aluminum alloy. The aluminum is injected into a steel die at over 5000 psi, assuring a perfect bond to the lascomite sleeve. Lascomite is a high tensile chrome-moly alloy, tougher and less brittle than cast iron. The 3.78 inch Nitralloy cylinders are cast with additional fin area in a symmetrical pattern to reduce thermal distortion. These cylinders are stock height so studs, pushrods and exhaust pipes of stock dimension can be used. Matching "zero

A small sampling of the pistons available from Axtell. Most aftermarket pistons are forged aluminum though cast pistons are probably good enough for nearly any street engine.

Axtell believes the cast iron cylinders make a better product than aluminum because they retain their shape when hot and do not "grow" like the aluminum cylinders do.

clearance" forged pistons with cam and barrel shaped skirt configurations are available.

S&S Cycle

S&S cylinders, made from cast aluminum with cast iron liners, are available with or without matching pistons. These cylinders come in various lengths to work with strokes from 4-1/4 to 5 inches. Stroke must be considered when ordering pistons as well, so there is adequate clearance at the bottom of the piston for flywheels and the other piston. S&S also has a new 4 inch aluminum cylinder meant for street use.

Pistons

Pistons come in various shapes, manufactured from either cast or forged aluminum. Forged aluminum is typically much stronger than cast aluminum, thus we all think we need forged pistons. The forged or cast controversy is another instance where things aren't as simple as they seem.

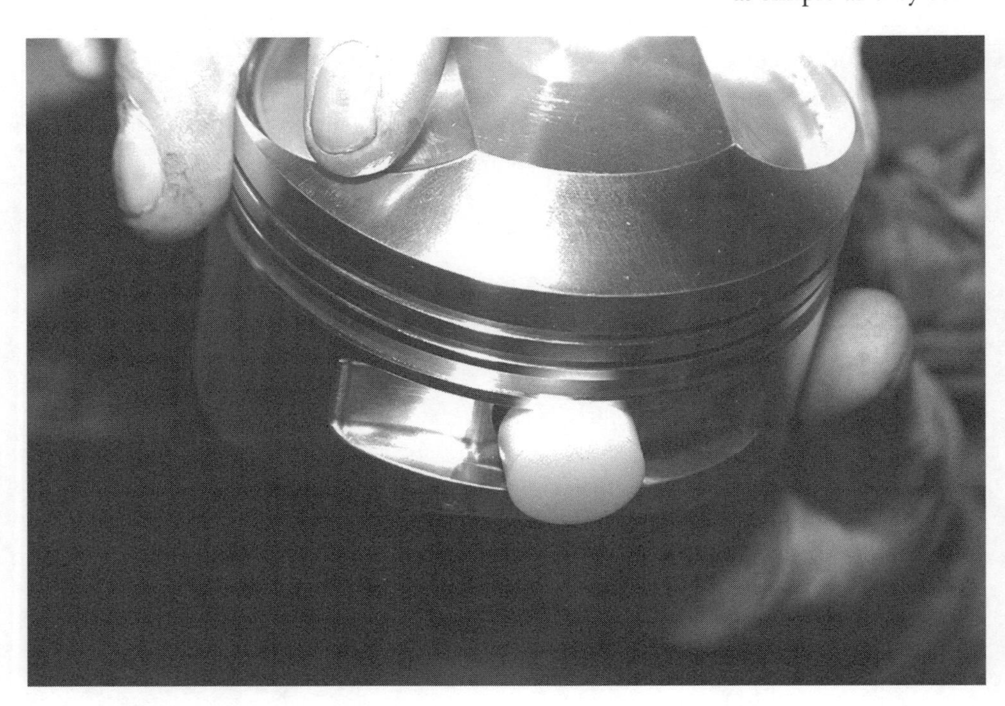

Some of their high performance pistons use a Teflon button to keep the piston pin in place instead of circlips.

Up until a few years ago forged pistons had to be fit somewhat loosely, in anticipation of the fact that the pistons would expand as they warmed up. Cast pistons on the other hand were made from a different aluminum alloy, one with a high silicone content to slow down the growth rate. Thus the cast pistons could be fit more tightly than forged.

The new crop of forged pistons contain silicone, however. Thus the growth rate for these new forged slugs is nearly the same as with cast pistons. Some builders and manufacturers, however, still prefer a cast piston. One engineer I spoke with said the manufacturing process for cast pistons allows them to have a more complex shape, one that fits

Axtell sells their cast iron cylinders in various bore sizes with or without pistons. Mid-USA

snugly into a cold cylinder, yet expands in such a way as to maintain the tight fit and quiet operation when the engine warms up.

For a look at the real world of engine building I asked Lee Wickstrom, long time engine builder for Kokesh MC in suburban Minneapolis, how he chooses between a forged or cast piston. "I look for a piston that will give me the compression ratio I want in the engine I'm building," responded Lee. "I think the cast pistons are plenty strong for anything except a nitrous bike. I fit the pistons according to manufacturer's instructions. They do tell you to fit the forged pistons a little looser, but once the engine warms up the running clearance is the same as with a cast piston so they really aren't any noisier."

We offer no Piston Buyers' Guide, there are just too many to chose from. Most buyers will get their pistons directly from the cylinder manufacturer, in many cases as part of a kit with the pistons already fit to the cylinder. When you buy pistons be sure you consider the heads as well, so the compression ratio comes out correctly. And remember that some cylinder heads work best when used with a certain piston shape (see Chapter Three for more on this).

If you don't get the cylinders and pistons at the same time, at least consult the cylinder manufacturer for a recommendation as to the best piston to run in that particular cylinder.

Axtell Sales

Though it seems all the aftermarket parts for both cars and bikes come out of California there is in fact a tremendous amount of activity in the midwest. Axtell Sales of Des Moines, Iowa is one small company busily manufacturing and selling parts designed to help you build bigger, faster V-Twins.

Best known as the manufacturer of big-bore cast iron and ductile iron cylinders, Axtell also sells a tremendous variety of pistons and other engine parts. In addition to

In addition to their cast iron cylinders, Axtell manufactures a range of ductile iron cylinders for competition applications.

manufacturing and sales they also do complete engine building and tuning with the help of their in-shop dyno.

Rick Vandehaar has been with Axtell for 16 years and explains that Axtell is happy to assist customers looking for more cubic inches but who aren't sure exactly how to proceed, or don't know which combinations of parts work best.

"People need to tell me what they're looking for and then I can suggest a package," explains Rick. "Most riders want torque and power down low so they get that thrust right off idle. Some guys, though, want to go out and beat Japanese bikes, and they are going to need a very different engine than the one that has the power real low in the rpm range."

Rick explained that one of their most popular packages is a 97 cubic inch engine based on a 4-1/4 inch stroke and a pair of their own cylinders with a bore size of 3-13/16 inch. "We like to see a shorter stroke, as close to a square motor as possible. Those motors are fast, they rev quicker and have good longevity. I usually recommend a flat-top piston for

In addition to cylinders and pistons, Axtell sells complete engines. One of their most popular combinations is a 97 cubic inch unit based on a set of STD or Delkron cases, with 4-1/4 inch flywheels and Axtell 3-13/16 bore cylinders. The heads on this motor are Performer rpm castings from Edelbrock fed by a Mikuni carburetor.

those engines. That gives the owner about 9.5 or 10:1 compression. This is normally what the customer will need in order to work with one of the camshafts I suggest."

Rick goes on to explain that once the customer decides on the internal dimensions the next step is to choose a set of cases. At Axtell the cases commonly come from either STD or Delkron, usually the late-model style with crankcase breathing through the heads. The flywheel assembly is usually from S&S, and can be ordered already balanced as long as S&S knows the piston weight.

Some aftermarket cases have a raised deck, which means Rick must shorten the cylinders to match. The other important measurement is rod length. "We often use S&S Supreme rods," says Rick. "Or Carrillo eight inch rods for the bigger motors. The longer, eight inch rods help in a number of ways: They provide more piston-to-piston clearance at BDC (bottom dead center). And a longer rod with a better rod-to-stroke ratio works at a better angle, the pistons are pushing straighter on the crank so there's less side thrust on the pistons."

For the 97 cubic inch kits Axtell often uses the new Edelbrock Performer RPM heads. "Right now the standard head is a RPM from Edelbrock, but we use STD if the customer wants more performance," says Rick. "The STD head is more money because they come raw and must be set up, so you're paying for more labor. And we can machine the head gasket surface to take an O-ring or just leave them alone and use a copper head gasket."

The carburetor depends on the engine of course, but for their 97 cubic inch motors the crew at Axtell likes the Mikuni 45. For exhaust Rick reports that most customers chose a system according to the way it looks on the bike.

Before deciding on the camshaft Rick likes to sell the customer on the idea of using pushrods with 5/16 inch adjusters instead of the more common 1/4 inch units. According to Rick the difference in weight is negligible and the beefier pushrods flex much less which means the valves open when, and as far as, they are designed to.

He goes on to explain that, "Camshaft choices start with a Crane 550 for the average street rider. It's a good choice and will give good bottom-end power. If the customer wants a hotter cam, but one that still gives good bottom end power we recommend the Red Shift 654.

If he wants more top-end power, we usually install

the Red Shift 625 but there is some loss of bottom-end power. With the 654 we recommend some rocker arm geometry work. We use a Baisley roller rocker arm and valves that come with no keeper groove so we can determine the length that the valve stem needs to be to get ideal geometry."

The Crane HI 4 ignition is Axtell's spark-generator of choice, unless it's a full-race motor or the customer requests something else.

When asked, Rick explains that they don't like oil coolers, which could cause wet sumping of the motor. "We think the important thing is to get an adequate oil supply to the bottom end. With more horsepower there is more heat and you need to get that additional heat out of the motor." Rick recommends the installation of Baisley 2-to-1 gears on the oil pump because they deliver twice as much oil to the motor and effectively remove more heat.

Heavy wall piston pins are designed for heavy duty applications while the thinner material is fine for most street applications. You don't want the pin thicker than necessary due to the extra weight.

In terms of the mistakes he sees people make, Rick says that with a big bore people should consider longer connecting rods, "otherwise the piston rocks in the bore, loading and unloading the rings and shortening their life. I also see lots of detonation, because people don't use the right heat range plug or they've got too much compression. We like to see about 170 psi cranking pressure on street bikes. The other thing I see is people who don't realize that when they change the pipes they're probably going to have to re-jet the carburetor"

Connecting rods are made from different materials (forged or billet steel and even titanium for light weight) and different lengths. Longer rods help to keep the pistons farther apart at BDC and also puts less side thrust on the piston for longer ring and piston life.

RevTech has a variety of pistons available, including these cam-shaped, low-expansion cast aluminum examples which come in various diameters and dome shapes. Custom Chrome

INTERVIEW
SCOTT SJOVALL FROM S&S.

Scott Sjovall is a Research and Development engineer at S&S, and an ideal candidate to discuss the kinds of decisions we all face when trying to decide just how big, how fast and how expensive that new V-Twin needs to be.

Scott, can we start with your background?

I'm originally from South Dakota. I spent my high school and junior high years in Virginia, which is where I got into motorcycles. I worked at a little shop called Virginia Motorcycle Supply through my high school years. When I graduated from high school I moved back to Rapid City, South Dakota and went to school at South Dakota School of Mines for Engineering. During those five years I worked at the Black Hills Harley-Davidson dealership. I finished school in 1992. I've been here four years. Lately I've been working on the engine dyno. I'm also doing some special motor building and working with a break-in/longevity engine stand.

Do you answer some tech calls from customers?

Yes, during my first two years here I answered a lot of tech calls, until they started hiring additional technical people to answer those phone calls. Then I could work exclusively on research and development projects.

The Dyno has to be a great learning tool for somebody like you?

Yeah. I've learned a lot. I've always ridden - since I was 14 or 15 - and tinkered with my bike, but it really took me coming here and working on the Dyno and on the phone all the time to learn the ins and outs and mistakes that people make.

When people call you looking for advice, in terms of which

Complete engine assemblies are available from your local Harley-Davidson dealer. These are current production engines, essentially excess production made available by allocation to each dealer. Each of these engines has been test run and carries a factory warranty.

S&S motor to buy or which motor to build, what kind of parameters do you suggest they use. How do you help the customer find that one engine that's right for them?

A lot of it has to do with the type of riding they do and how much experience they have with a motorcycle. Does the guy want a bike he can still take everywhere and not worry about cold starts or longevity? If he wants to go touring and go places without worrying about what kind of gas he's going to get then we're going to want to go toward a milder, lower compression model. Something that is really trouble free. On the other end of the spectrum, if the guy is out trying to beat his buddies on Saturday night, then you start swinging the pendulum back toward the high compression 103 or larger inch motors. That's probably the toughest part of putting a motor together, figuring out where in the spectrum the rider really is.

When they call for advice, do you ask them about the kind of trade offs they want to make?

Yes. As far as engine life, vibration, start-ability, cost, maintenance on the motor. It really depends on what you want to do with the motor.

Take, for instance, if a bike is an older model Evo. The starter system on some of those bikes generally isn't that good. You don't want to stick a high compression 98 inch motor in that bike because: for one, it won't start. For two, it's a motor that's going to have to come down every so many thousand miles for maintenance, and the owner may not want to deal with that. So you're going to push him back toward something in-between, maybe a 89 or 96 cubic inch set up. Something that is still going to start and give good power, but a motor that he or she doesn't have to worry about.

Going back to what makes a good motor, why don't we start with the basics. What makes a good set of cases?

The most important part of the cases is keeping the dimensions correct. All your center lines and all the deck heights have to be perpendicular and dimensionally correct. That was probably the biggest problem we ran into with building motors before we got into the case business. We had to blueprint the cases before we

Some big-bore engines move the lifter bores to the outside, which necessities a longer pinion shaft. These engines still run the standard camshafts and valve gear.

could use them. Correct dimensions is probably *the* most important thing.

The second thing is to buy a set of cases that are strong enough so they won't break. Most of the companies have pinpointed the weak areas in the O.E.M. cases and fixed that. There are a lot of strong cases out there.

The third thing with cases is you want something that is going to fit in the motorcycle, with the correct exterior dimensions. Something you don't have to do a lot of grinding or chiseling on. Something that makes your life easier, the mechanic's life easier, and is therefore more cost effective to put in your motorcycle.

Your cases are cast aluminum?

Yes, 356 T-6, XL material. It's been doing real well for us. We've added strength in the areas between the bores - O.E.M. cases had a problem breaking out there and also had a problem spitting out the drive side inserts. We've addressed those areas. And we've changed some oil passage patterns inside the cases to really help out those weak spots, particularly the ones behind the tappet blocks.

Ours have been a good set of cases and we've had very few breakage problems. I haven't actually seen a cracked set of our cases yet. That pretty much is what you look for. You don't want overkill in the case design. You start getting too much weight where you don't need it. It looks tough but you're also taking away from your available horsepower. We really look for the best balance we can find between strength in the places we need it without making them look really massive.

So you think some of the really heavy duty, billet cases are overkill for the street?

Definitely. They look nice. They're also overkill for most people's pocketbooks. Our Dynos are real abusive to cases. I can generally shake an insert loose within 30 minutes, that's with a stock set of cases. That's an area where we've spent much of our development time. I believe the street cases we have for sale now, all the way up to the available bore sizes, can handle the amount of heat and abuse we throw at them.

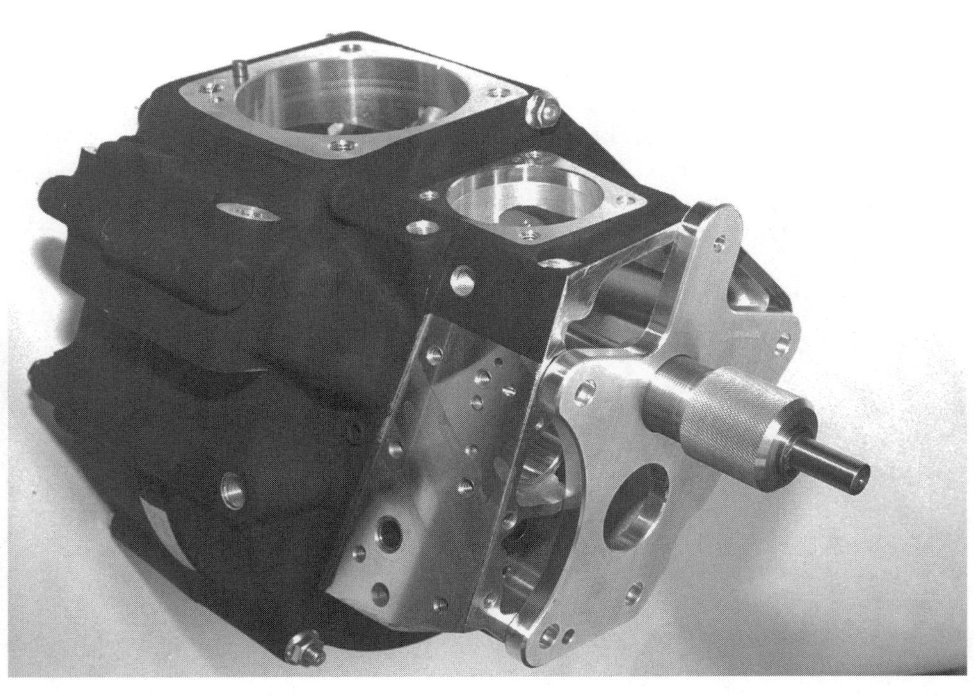

This special tool from Zipper's is driven by a drill and used to remove case material where it interferes with passage of the inner cam lobe.

High-lift cams sometimes contact the case as shown here. It's a good idea to check this and do any clearancing work before the gear case is completely assembled.

How big a bore size can we use with the current cases?

We like to use 3-13/16 inch as the maximum bore size. That fits in easily. As far as the strength of the cases, we've run a 4 inch bore real successfully, but it's a fine wire trying to get them in. You have to do some playing around. It's more a question of room than strength. (Editor's note: Cases that can accommodate larger bores will soon be available.)

What about pistons and cylinders. Again, with an eye toward high-performance street use for a typical rider, builder. Do you have to buy them in a set so that they're already fit and matched to one another? Are there any problems with regard to bad matching in terms of compatibility and that kind of thing?

You want to buy them so they would at least match each other in theory. If you have a company that specializes in building pistons for aluminum cylinders you kind of want to go with aluminum cylinders and vice versa. If you buy from a company specializing in steel cylinders, then they probably know what works best in their cylinders. We've got two different piston lines that we sell here. A steel cylinder market, which is the racing market for us. And we have a different piston line for our aluminum cylinders. I like to say that if you buy from a manufacturer and they match the set, then you don't have to worry about it. On the other hand, there are a lot of very good mechanics out there and capable shops, and if the particular set up you're looking for isn't available as a package, they probably wouldn't have a problem with buying the parts separately. Just be careful that everything is meant to go together.

Seen here is the crankcase venting (exiting the head) used on 1992 and later engines.

Late model (1992 and later) V-Twins use a different oil pump from earlier engines.

There are some new pistons out there on the market, the technology has really come a long way in the last few years. The newer high-silicone-content pistons that are available now hold their shape real well and work exceptionally well with aluminum cylinders because of their ability to retain their shape as they warm up. You can fit them up pretty tight in the cylinder and it gives you a lot longer life. If your machine shop or mechanic isn't up to date with some of the

These Sputhe cases are cast from 356A T6 aluminum and feature extra case bolts and increased wall thickness for heavy duty applications and bore sizes of 4 inches and more. Arlen Ness Inc.

things going on, you could run into problems there.

I asked one of your other engineers about the cast versus forged piston controversy, and he said that in the last couple years, there are alloys in the forged line that can be fit tighter than those in the old days. Are the high-silicone-content pistons you described, are those forged pistons?

Yes, those are forged. That is all we sell for the Evolution line, because we've had such good luck with them. There was never a need to go to a cast piston for a better fit.

Let's talk about cylinder heads. What goes into the design of a good set of heads. And as long as we're talking about heads, why does S&S insist on using their own pistons with a set of their heads?

Our design criteria called for a high quench combustion chamber and fairly high compression. So our piston was designed to work with that chamber. It's a very close tolerance chamber, about a .045 inch squish area all the way around. It's not a flat squish area, like the stock ones, it actually fits up into the dome. That's the reason you have to use our pistons with our heads. The reasons we went with that design is it was the only way we could keep the flow up in the ports and get the quench where we wanted it. Also, by changing pistons it allows the same head to be used with different displacement engines.

These big bore cylinders from Sputhe can be ordered in a kit with pistons or mated to a set of Sputhe cases and 4-1/4 inch flywheels to create a 95 cubic inch V-Twins. Arlen Ness Inc.

The burn pattern was something we worked with a lot also. That baseball-diamond type shape has a lot to do with controlling detonation and getting a better burn. We also had to consider the amount of turbulence we wanted in the chamber. That all came into play.

Do you want to define the term quench area?

The quench is an area in the cylinder head where the

piston and the cylinder head, at TDC, are in close proximity to each other. In our case, it's a .045 inch clearance area. And what that area does is force the fuel/air mixture in from the outside of the piston or where the quench area is, in toward the spark plug or the middle of the chamber. It creates a lot of turbulence. It gets the gases emulsified with the air so you don't have a lot of fuel droplets in there. That's what the quench does. To a point the more quench you've got, the better, because the more turbulence you're going to have. You can pretty much start the flame in a smaller area of the chamber. Basically, there's a lot of different ideas of where you want the quench and where you want it aimed at. We call ours a closed quench chamber because it goes all the way around.

What if I'm just a street rider, what do I look for in a set of heads? What are my criteria. Since there's now so many aftermarket heads out there, obviously you tout your own brand and that's fine, but from a purist's standpoint or an engineer's standpoint, what goes into a good set of heads? What do I look for when I'm buying heads.

It kind of goes along the same lines as the cases, where you want to make sure all your clearances are set up correctly. You want a valve train that's going to hold up for a long time. That's going to depend on the materials that the valves, seats, guides and the heads themselves are made out of. Because the heads get really hot and take a lot of abuse. They have to be made out of a good material that is not going to lose its strength. Look at what materials the head and all the associated parts are made out of. You have to make sure that anything else you want to add to the bike is going to fit on the motorcycle. You don't want a set of heads that aren't going to allow you to use a certain exhaust system or a certain carburetor that you may want to use down the road. So fit is an important question.

The other thing you want to think about is rideability, startability. You would find that out from your chamber design. How high is the compression ratio? Is it high enough to give you the horsepower you want? Is it too high where you get hard starting and detonation? That in turn also leads to port design. Everybody knows that generally a bigger port is going to flow more. But that's not all you have to worry about here. More air flow isn't necessarily good if you don't have velocity to back that up. That's when those big ports

Cylinder heads should be chosen according to your budget, riding style and the other components used to assemble the engine.

Most engine kits do not include the cam cover - though S&S now makes this cover of their own.

will get you into trouble. I don't know of any really bad heads on the market, but some are better than others.

You can have too big a port in the sense that you lose too much velocity?

Definitely. That's why we market two different heads. We have a race head, which flows big numbers and gets a lot of air in and out of the motor. But on a street bike, it's just really slow and sluggish. It doesn't make the horsepower down low. So it's a trade off. What heads you want to go with - where do you want your power? Try to get the best combination. If you do it right, you can get a good set of heads that have high velocities through the ports. That's what we try to work with. It's the reason it took so long to come up with our heads. You also want to look at parts that are available in the future for the heads. If you buy a set of heads and everything on them is one-off, will the company still be around in five years and will you be able to buy parts for the heads?

Most cast engine cases are finished on elaborate CNC type machining centers like this one at S&S.

Again there are billet heads on the market. They're kind of high end. Is there a significant advantage to using a billet head from a material or design standpoint?

Not that we've seen here. The only instance that I can think of where one of our cast heads wasn't strong enough for the application was in a supercharged nitro bike at 200 mph, and even at that, we just changed the casting a little and it went on to be the first Harley to go over 200 mph. Billet heads are nice, they're a very exact head. They're a pretty head. And if they do it right, they can keep a real close eye on the ports. They are generally a good head. But as you pointed out the price is higher and many don't flow as well as our cast heads.

As long as we're on heads, once I decide I'm going to choose

Brand X, how do I choose a good carburetor and pipes to go with that. The market is going more toward kits, but if I don't use a kit, how do I put together my own combination of parts?

You have to pay a lot of attention to where in the rpm range you want your horsepower. More restrictive exhaust with smaller carburetors bring out the bottom end. On the other side, those with the big carburetors and a set of drag pipes, you'll have really good 6,000 rpm horsepower, but your throttle response will be very soft.

There are very few pipes on the market, that we've tested, that cover the whole spectrum. S&S has always favored the stock header system with a set of flow-through mufflers. Lately we've been doing a lot of work with the Thunder Header. SuperTrapp is real good for mid range and low-end performance, though it gives up a little bit on top. B.U.B. Step-mothers work well on big inch motors. That's what you want to think about. Everybody thinks drag pipes are fast, what you need is a good exhaust that's matched to your riding style. And a carburetor that's matched to the rpm range and the size of your motor. For example, our Super E is better than the G for heavy bikes that run at lower rpm.

Generally, I would talk to a manufacturer and see what they recommend for a particular motor. They can tell you what's going to work best for you. A Sunday afternoon driver with a sidecar is not going to want the same motor as the guy who is trying to win the Dyno shoot out.

Do you have any comments regarding mistakes people make when they go looking for horsepower. Or mistakes they make when they find high performance parts?

I would like to stress that if they're looking at horsepower numbers, they should look at the numbers in the rpm range where they are riding. Everybody out there, and us included, offers the horsepower numbers that the motors will make. That's great - it tells you the potential of the motor. But if you're riding a bike around 2,500 or 3,500 rpm look at the torque numbers down there. That's what makes a bike fun to ride.

Really keep it straight in your head. Don't talk horsepower numbers, at least not at high rpms. Talk torque and lower rpm horsepower numbers right where you want it. Right where you think the motor is going to perform best for you. Most of the calls we get from people who aren't happy with their motor combinations have a high compression ratio, a big cam, a big carburetor, and a set of pipes that sound good. But the engine doesn't make usable power lower in the rpm range.

There is a flip side to this. There are people out there that bought our kits with the whole intention of running the bike on the Dyno. Had we known that when they ordered the parts we could gear things that way. But a bike that turns 100 horsepower on the Dyno isn't nearly as much fun to ride on the street as the 75 horsepower bike with the "bolt-in" cam and restrictive mufflers. It really depends on what you want to do. Look at the numbers. A heavy bike or a touring bike doesn't need the 103 inch motor with an aggressively timed, high-lift cam that might work good in a hot rod FXR.

Scott Sjovall from S&S has been involved with motorcycles since he started riding in high school.

Chapter Three

Getting A Head

Two valve, four valve, cast or billet

Introduction

There are plenty of heads out there, from used factory castings to billet four-valve designs. At least one builder interviewed for this book suggests that you chose the heads first, not last, and then make sure everything else (pistons, cylinders, and all the rest) are compatible with the heads.

Your new heads have to match the stud pattern for the engine. As mentioned earlier, most bore sizes up to 3-13/16 use a standardized stud pattern. Bore sizes of

Cylinder heads come in every imaginable shape and configuration. These competition heads with the unusual port shapes were designed by Craig Walters.

four inches and larger, however, use one of a number of stud patterns.

The heads must be matched to the pistons. Some heads work fine with a flat-top piston while others need a domed piston that fits finger-and-glove into the recess of the cylinder head. Be sure to ask the head manufacturer for a recommendation in regards to the pistons you use - which will best compliment the combustion chamber shape and which combination will provide the compression ratio you've decided upon.

If you are relying heavily on your local V-Twin engine builder for advice, then you might want to take that advice when it comes to choosing the heads. If you're assembling a "known combination" of parts, then it doesn't make much sense to deviate from that plan.

Consider that among all the new companies bringing heads to the market, not all will be here in three or five years. Try to buy from a company with a good reputation, one that puts good quality components in their heads. A company that will be there the next time you pull the engine apart for a rebuild or further improvements.

Remember that it is sometimes cheaper in the long run to buy a more expensive set of heads than it is to buy cheap and the pay a high per-hour figure to have them ported to flow more air. (see Shop Tour for more on porting). And while we all want heads with good flow numbers, there is such a thing as too big a port on a head meant for street use. A final note on flow, be sure to ask the manufacturer for their flow charts before paying for the new trick cylinder heads.

While dual-plugging the heads adds to the sex appeal and in some cases the horsepower, it isn't necessary on most hot street engines. In fact, some head manufacturers recommend against the installation of the second spark plug.

CYLINDER HEAD BUYERS' GUIDE

Edelbrock

Long known for their automotive parts, Edelbrock has entered the V-Twin market with a pair of cast heads intended to offer increased performance for V-twins.

The first head they introduced, the Performer, comes with a 1.850 inch intake and 1.610 inch exhaust valves, a D-shaped exhaust port (which accepts all stock-style exhaust systems) and rectangular-shaped intake port.

More recently Edelbrock has introduced a second head known as the Performer RPM head. This second model comes with larger, 1.94 inch and 1.625 inch valves, and "CNC ported" passages. The Performer RPM is available in two combustion chamber sizes and comes with high-test springs and titanium retainers, designed to allow the use of camshafts with a lift up to .650 inches. If you would rather do the finish work on the valves yourself the Performer RPM heads are available in a "raw" configuration.

Feuling 4-valve

It was only a matter of time before 4-valve heads, now used on many high-performance automobiles (and some Grand Prix cars and bikes early in this century)

RevTech heads are cast from 356 aluminum and come either finished or in raw form. All RevTech heads feature raised intake ports, bathtub-shaped combustion chamber and standard exhaust flange configuration. These raw heads will accept seats for oversize 1.940 inch intake valves and is already tapped for dual plugs.

Checking For Valve To Valve Interference

If your heads have larger valves or new seats installed or if a new performance cam has been installed, being able to easily check for possible valve to valve interference will be helpful. For all H/D heads it is possible to do a simple calculation to see if valve to valve interference might be a problem which will need attention.

1. Andrews Products lists valve lifts at TDC (Top Dead Center) on all cam instruction sheets. Write down the number for your cam. For an EV59 cam, the TDC lift = .233 inches (see data on page 7 of the catalog).
2. Minimum valve to valve clearance should be .060 inches.
3. Calculate the minimum valve separation distance as follows: minimum Valve Separation Distance = TDC lift + clearance.
4. For EV59 cams, Minimum Valve Sep. Dist. = .233 + .060 = .293.
5. Measure the minimum separation between the two valves **when they are seated** (as in diagram). If actual measurement is not at least .293 inches, modification will be necessary to avoid valve to valve interference. (Cut seats deeper or back cut valves).
6. Remember, this technique is NOT for piston to valve clearance.

Used with permission from the Andrews Products catalog.

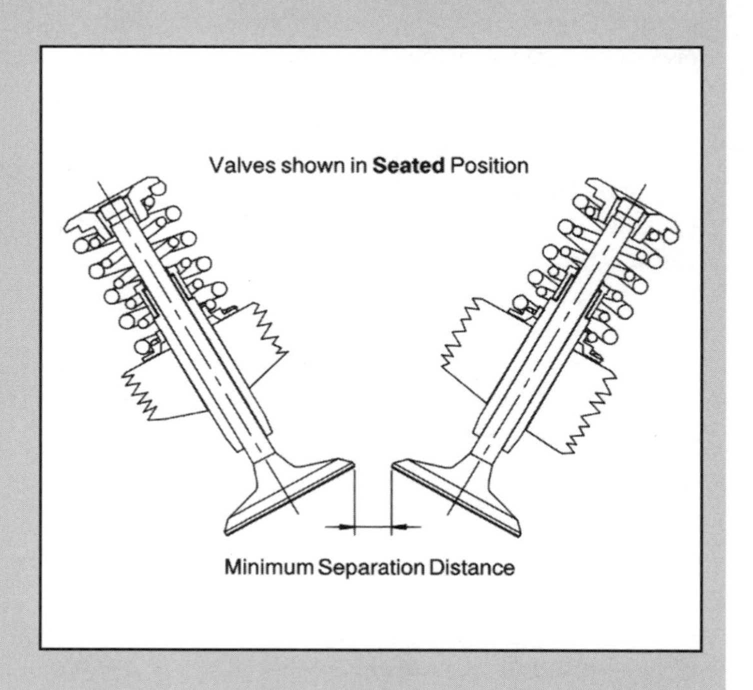

would "cross over" to the land of V-Twins. In essence, two small holes can be made to pass more air than one big one. By using two smaller ports total flow can often be increased (especially at modest valve lifts) while still maintaining good velocity through the port.

Fueling cast heads come in various stages, starting with a kit that uses the stock carburetor and Feuling anti-reversion exhaust pipes with the 4-valve heads. Stage II kits include a dual-carb intake, so you can hang one carburetor on either side of the bike, and a compatible high-performance camshaft. For the truly serious horsepower junkies Fueling offers stages III through V.

Johnson Performance Engineering

One of the "New Kids on the Block," Alan Johnson usually spends his time manufacturing billet heads for Top Fuel cars. That same expertise has now been used to manufacture a complete V-Twin head which is shaped by CNC machines from a raw chunk of 6061 T6 aluminum.

These new heads come with their own intake manifold and accept stock and aftermarket two-bolt exhaust systems. Designed for single spark plug use, these heads are available from Johnson Engineering.

Merch Performance Inc.

From Canada comes a cast head with CNC machined bathtub-shaped combustion chambers. A raised intake port helps to increase flow, all the way to 150 C.F.M. at 10 inches of water with the 1.900 inch intake valves at .500 lift (without any additional porting work). Merch heads can be ordered with bigger valves, four-bolt exhaust and early or late-style breathing. You can even order Merch heads to fit bores of four inches and more.

Patrick Racing

Billet aluminum heads from Nigel Patrick come in various stages of "high performance." The basic head features a 76cc combustion chamber and 1.94 and 1.60 inch intake and exhaust valves. Options include non-stock valve angles and some very large diameter valves - all the way to 2.25 inch for intakes and 2.00 inch for exhausts. Though Patrick makes heads for your stock Evo, the polished billet heads are more commonly seen atop a large bore motor like a Keck.

RevTech

One of the first large aftermarket companies to

offer a complete, ready-to-install, cylinder head, RevTech now offers their heads in both stock and big-bore configurations; in stock and high compression models.

These aluminum heads are cast from 356 alloy, heat treated to T6. The stainless steel valves measure 1.940 and 1.610 inch respectively for the intake and exhaust. Though the intake ports have been raised from the factory location these heads will accept all stock intake manifolds. 3-piece valve springs and heavy duty retainers allow the use of camshafts with lifts up to .600 inches. RevTech heads can also be ordered "raw" to be finished by the engine builder of your choice.

Rivera Engineering

Rivera Engineering offers for sale two versions of their billet 4-valve heads, one designed for bore sizes up to 3-13/16 and another for monster motors with a bore up to 4-1/4 inch.

Designed as a bolt-on improvement for most V-Twins, these heads accept stock intake manifolds and exhaust pipes. Access to the centrally located spark plug is said to be much improved when compared to earlier cast 4-valve heads offered by Rivera.

S&S

These cast aluminum cylinder heads feature a 2 inch intake and a 1.610 inch exhaust valve. Both the intake and exhaust ports use a unique shape, the intake in particular uses a vertical vane intended to help route the incoming charge around the intake valve.

S&S heads come with their own manifold, one with a horizontal vane meant to create controlled turbulence and aid in complete burning. The exhaust ports use standard mounting dimensions so stock and aftermarket exhaust systems will bolt on without modification. The combustion chamber is a closed, high-quench chamber which necessitates the use of S&S pistons.

S&S heads do not come taped for dual plugs, and S&S feels "The superior mixture flow and turbulence also provide excellent flame travel..... dual spark plug configurations for normal operation are not needed or recommended."

STD

This cast head with the "bathtub" chamber normally comes in a raw configuration. That way your engine builder (or STD) can install seats and valves matched to the engine size and use, and add any porting work they feel is necessary. These high performance heads feature a raised intake port for improved port shape, and a four-bolt pattern on the exhaust port so either a stock or flanged style exhaust can be used. Despite the raised intake port, these heads will accept stock intake manifolds. These heads are available in either a single-plug or dual plug configuration.

In addition to their raw head castings with the bathtub combustion chamber, STD is currently unveiling a new, ready-to-wear, cylinder head. This new head will come completely assembled with 1.840 inch intake and 1.610 inch exhaust valves.

Sputhe

Sputhe cast heads come in two variations, designed for 80 inch V-Twins, or the larger version for 95 cubic inch and larger engines with bore sizes of 3-13/16. These heads feature a large fin area for better cooling and raised intake ports for improved flow. Buyers can choose between aluminum bronze or manganese-meehanite valve guides. Intake valves are one-piece stainless forgings while the exhaust valves are titanium-nitride coated one piece austenitic forgings. Valve springs can be ordered to best match the acceleration and lift of your camshaft.

Before You Bolt On The Heads

Some things to consider

Whether the heads you are bolting to that new V-Twin are heavily modified castings from a company like STD, or nearly-stock units from The Motor Company, there are a few things to consider before you bolt them on. Your list of possible concerns includes:

1. Possible valve to valve clearance problems. To quote Lee Wickstrom from Kokesh MC, "As you get into wilder cams and heads with bigger and bigger

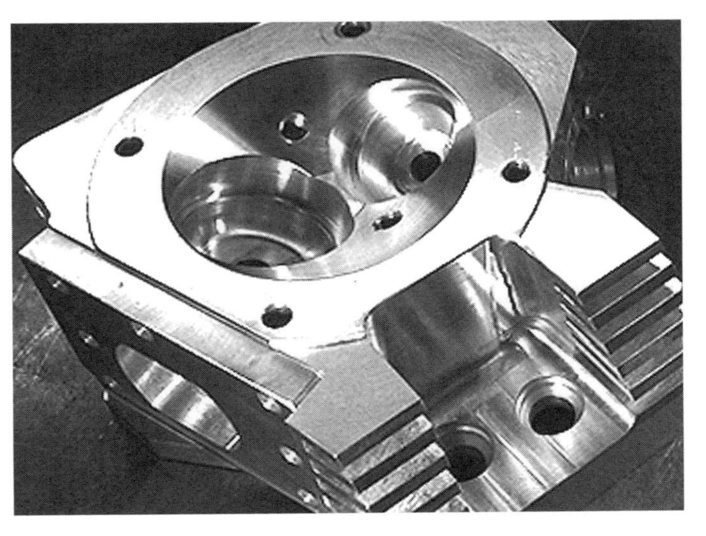

Monster motors need Megaheads. These billet heads from Hyperformance are designed to accommodate bore sizes up to 5-1/8 inches. Valves measure 2.6 inches on the intake side and 2 inches for the exhaust.

valves, you need to pay more attention to this matter." (See the material from the Andrews Catalog included in Chapter Four for more on this potential problem.)

2. Possible valve spring coil bind. You need to know the dimension of the spring when it "stacks." Now, with the head assembled (with a light spring) and the valve depressed to "maximum lift" compare the dimension between the spring retainers to the coil-bind figure. Most cam and valve-gear manufacturers recommend that you have .050 to .060 extra clearance between maximum lift and coil bind. It's obvious that you shouldn't lift the valve to the point where the spring hits coil bind. Less obvious is the fact that if your cam/head/spring combination has the spring *near* it's coil-bind limit at each camshaft rotation the springs will soon fatigue.

After the valve job and porting have been done, Dan strips the head and reassembles it with "light" valve springs.

3. With the valves at maximum lift, you also need to ensure that there is at least .050 inches of clearance between the top of the valve guide/seal and the bottom of the upper spring retainer.

4. Though you need clearance before coil bind, you also need a spring with enough pressure to keep the lifter rollers following the cam lobes at high rpm. Some builders like to use a set of springs matched to the camshaft, but even then you need to check for coil bind and clearance to the top of the valve guide.

5. Valve to piston clearance. Though it sounds like too much extra work, professional builders recommend assembling the engine with clay on the valve pockets of the pistons, rotating the engine by hand a few times and then disassembling to make sure the clay is at least .080 inches thick where the valve came closest to the piston.

Note: Much of this is covered again in the Zipper's Shop Tour.

After reassembling the head with light springs, Dan measures the installed height.

If there isn't enough spring pressure spacers can be used below the spring - but each spacer moves the spring closer to its stacked dimension.

SET UP A HEAD AT ZIPPER'S

This work is done at Zipper's Cycle in Jessup, Maryland

Dan Fitzmaurice, owner of Zipper's, starts the head assembly with a bare head. At this point the valve job and porting work are already done.

Before starting on the assembly, Dan provides a brief discussion of valve springs. "The valve springs must be a good match for the cam. In Zipper's opinion, lots of Harley-Davidson engines are assembled without enough spring pressure. Too much spring pressure is not necessarily as hard on the valve train as people might think. There needs to be a balance here because out-of-control valves are harder on the valve train than high spring pressure. Get a spring pressure recommendation from the camshaft manufacturer or from the local cylinder head expert."

Dan starts by assembling the head with a light spring installed in place of the real valve springs. Now he can check the installed height. With the Baisley machine he can compress the spring to the installed dimension and check the spring pressure.

If the pressure is too light, a shim can be placed under the spring, but you can only add so many shims before you reduce the amount of travel available to the spring before it "coil binds." In some cases you might have to use a different spring to achieve the correct pressure and travel at a given dimension.

To check for coil bind, Dan compresses the outer spring, with upper and lower collars (and any spacers) in place and records that dimension. (Warning: A compressed spring is kind of like a time bomb. Be sure it doesn't detonate - get loose - while you have it compressed.)

Because he knows the installed height, Dan can make sure that the total valve lift plus a clearance factor won't compress the spring to the point of binding. The Andrews Camshaft people feel

This machine from Baisley is used to check the seat pressure of the installed spring.

you should have .060 as "clearance" before spring bind occurs. Minimum spring height should equal the stacked height plus the .060 inch clearance plus the maximum lift of the camshaft.

Note: it's a good idea to check installed pressure and clearance before coil bind even with springs that were shipped as part of a kit - because the manufacturer of the spring has no idea which set of heads you own or exactly how the valve seat is positioned in the head (all of which is affected by any porting work and valve jobs that have been performed on the head).

Dan explains that you also need to be sure that the upper spring collar doesn't contact the top of the valve guide or seal at total lift. To check this he compresses the valve the full amount (with the light spring in place) and checks the clearance between the bottom of the upper spring retainer and the top of the guide and seal.

With regard to possible valve-to-valve clearance (which should be checked before the valve job) follow the Andrews' procedures outlined earlier in this chapter.

Interview: Craig Walters, Walters Technology

Craig Walters of is not only a legend in drag racing circles, but also the man who brought us the Accelerator computer program designed to help both amateurs and professionals predict the outcome, in horsepower and torque, of a certain combination of parts. If that weren't enough, he's also very busy working behind the scenes helping the various aftermarket companies develop heads, camshafts and carburetors.

Craig, how did you get started designing parts for hot rod Harley-Davidsons?

We started out drag racing Sportsters. We blew up the first one, but the second one ran pretty well. This would be about 20 years ago. The bike was a Shov-ster, a Sportster with Shovelhead jugs and heads. We did a lot of work to those heads. We welded up the ports and made the combustion chamber smaller, it became very similar to the evolution heads we have now.

Could we talk a little bit about your more recent work in the after-market? Like the work you did for S&S?

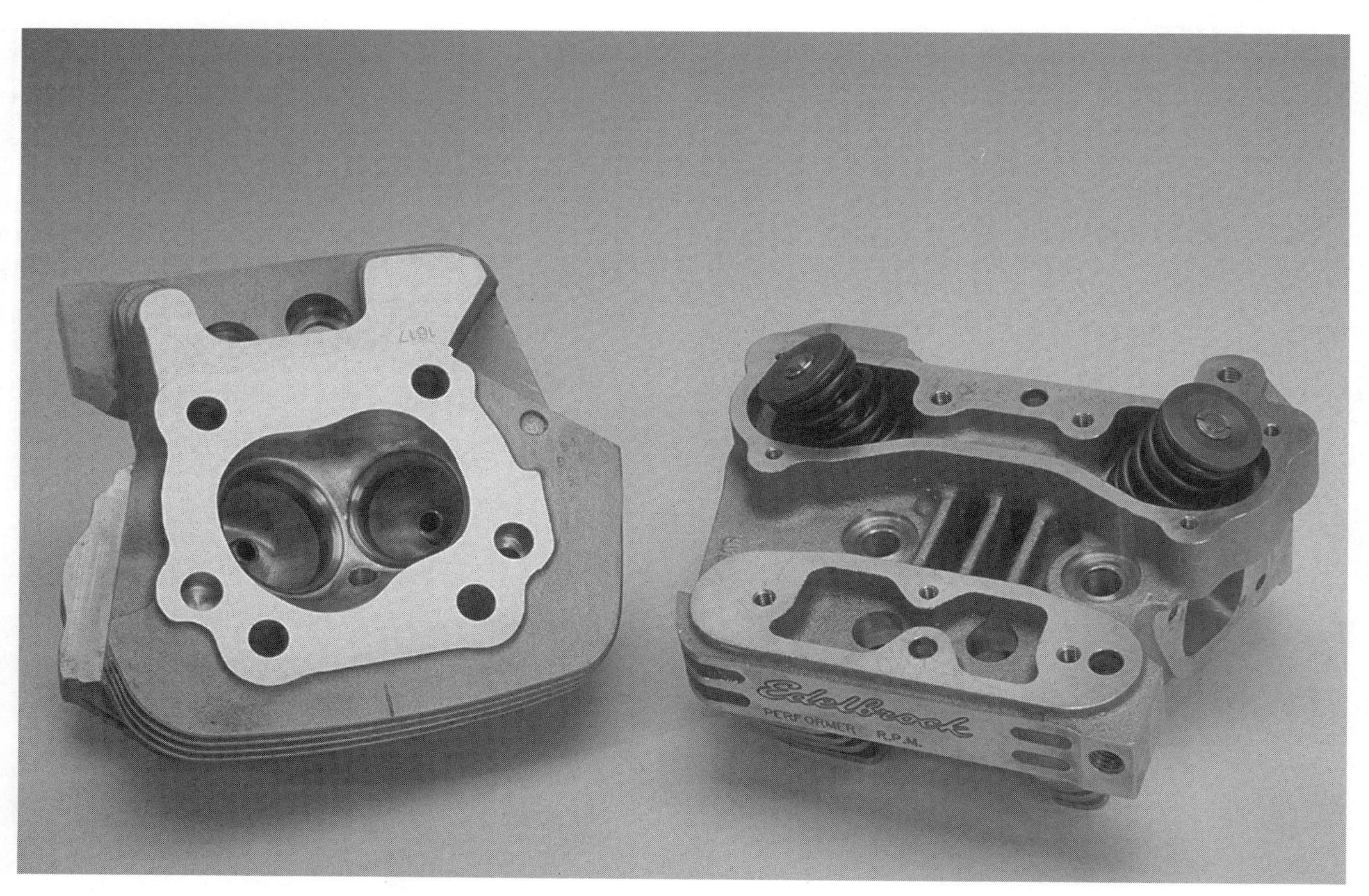

Edelbrock heads come in two models: Performer and Performer RPM. The "RPM" head comes with larger 1.940 and 1.625 inch intake and exhaust valves.

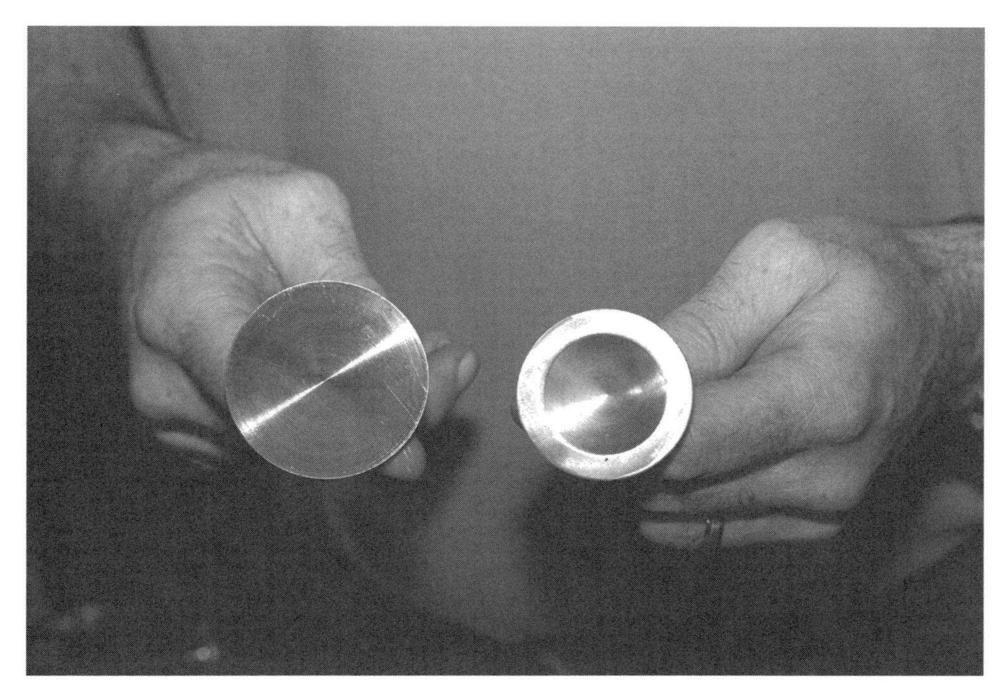

reputable head guy who has done heads for that type of application. The biggest thing to look out for, and you have to be truthful with most head guys, is what you're going to be doing with the bike. Is it going to be a hot rod tavern-to-tavern bike? Is it going to be a touring bike? That is probably the biggest thing I've seen in this industry. Nobody tells us their actual intent so we can't provide them with the best heads.

It's a misconception in people's mind that they have to spend a ton of money before they get anything good.

In setting up a head, more than just the valve size needs to be considered. Different shapes (as shown) affect compression ratio. Weight, shape on the backside (which affects flow) and the material the valve is made from are all important factors.

At one point I had a company out on the West Coast, we did quite a few street heads, about 500 pairs a year. We found the need for a good head. The only kind being cast at that time was STD's, and they still required a lot of work. So when S&S wanted to cast a head they hired me to do the project

And I've worked for other companies too, but I'm under disclosure not to disclose sources. I can say we have done something for every company out there one way or the other.

How long have you been working on the accelerator program?

12 years now. We started using it at the drag strip to generate track numbers, horsepower, and airflows. Now we've programmed it to be user-friendly. That conversion took a year and a half.

Let's jump into building a motor. We're going to talk about heads, and my emphasis is on high performance street stuff. So I'm building this motor from scratch. Here's the big question, "How do I chose the heads? What are the parameters that I need to consider?"

You need to know how big the motor is. How much air flow it's going to need. The only way to really determine air flow is to either buy our Accelerator program and it'll give you an idea or get a hold of a

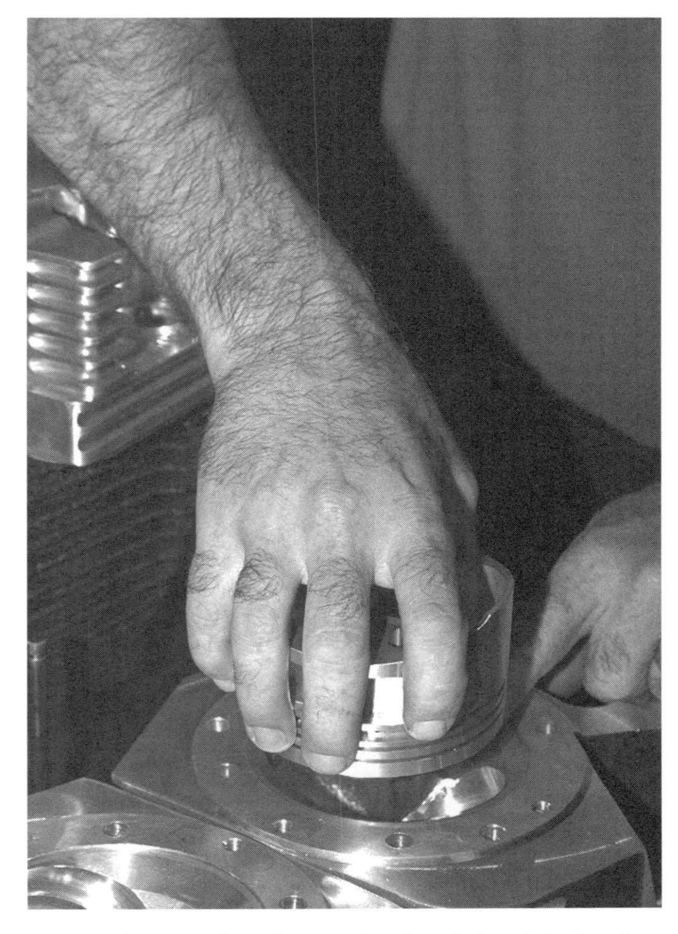

Piston shape needs to be considered in light of the head design. You want the correct compression ratio and a dome shape that works in harmony with the head's combustion chamber shape.

A pair of new, factory D-shaped heads with stock ports. Lee uses an intake valve (all the demo work is done on the intake valve) that is notched to better determine where the sealing area is relative to the edge of the valve.

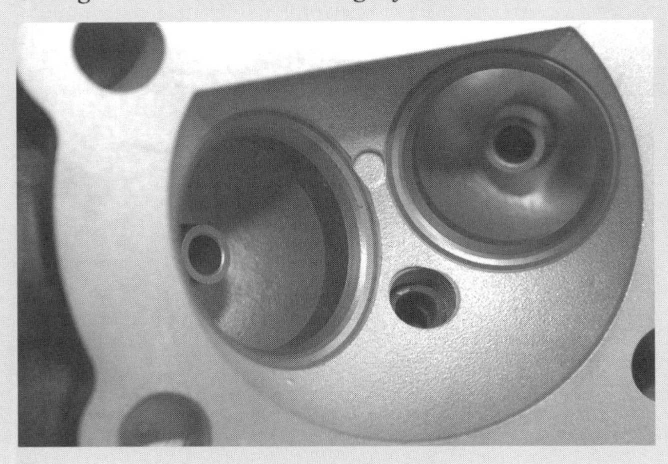

This close up shows the stock seat and port area before the new valve job and porting work.

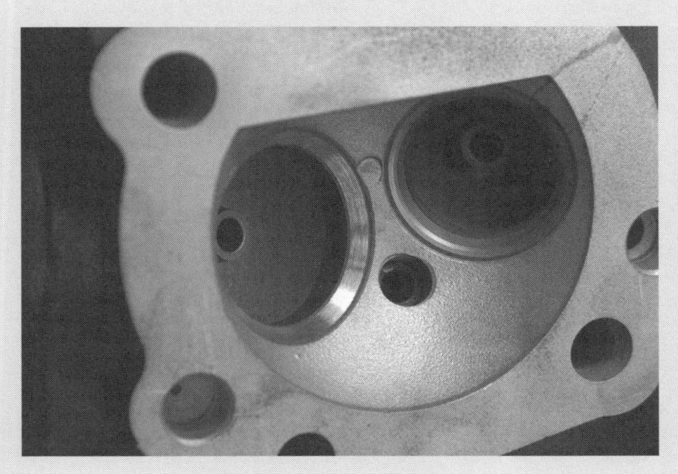

The first step in porting the head is a good three-angle valve job (a five-angle would work just as well). And the first part of the valve job is to "flatten" the seat area with a 45 degree stone.

SHOP TOUR: PORT A HEAD AT FRITZ T. WILSON RACING

Lee Wickstrom works as a mechanic at Kokesh Motorcycle Parts in Spring Lake Park, Minnesota. In the evenings, however, you can find Lee working in a different shop - the Fritz T. Wilson Racing shop. Fritz T. is the name of Lee's own small high performance company and much of the work Lee does "after hours" is porting work on a variety of cylinder heads. Lee recently worked through a porting operation on a set of factory Harley-Davidson heads, taking breaks so I could photograph the proceedings (though the changes that take place in a port are subtle and hard to document with photographs).

Where the changes in port shape really show up is in the increases in air flow as documented on a flow bench. When we finally finished sticking the camera lens down into the intake ports we took camera and notebook into the "office" where Lee keeps the Superflow 110 flow bench. What follows is a description of what goes into a good porting job and how the improvements are documented.

HEAD PORTING 101

A good porting job starts with a good valve job. Lee wants the actual seat (the areas where the valve seals against the "seat") to be as close to the outer edge of the valve as possible, because this in effect makes for a larger diameter valve and port. In the case of these factory heads the actual sealing area is too far down - about in the middle of the valve's 45 degree surface. Lee moves the sealing surface closer to the edge of the valve in a series of cuts:

1. First a 45 degree stone is used to eliminate the current sealing area.

2. Then he uses a 30 degree cutting tool to define the upper edge of sealing area.

3. Next comes a 60 degree cutting tool to define the lower edge and the width of the sealing area. Note: For street use Lee likes to see a sealing area .040 inches wide for the intake and .060 for the exhaust seat. To gauge the width of the intake seat he uses a piece of wire known to be .040 inches in diameter.

4. Lee finishes by lapping the valve into the seat with valve lapping abrasive. The lapping is a good way to finish off the valve job (it eliminates any high spots that might exist on the seat or valve) and it also leaves an imprint of the seat on the valve making it possible to see exactly how close to the edge the sealing surface actually is.

5. Lee goes one step further and holds a light above

the combustion chamber while looking into the port for light leaks.

Now we can start on what most people think of as the porting work and as you may or may not know most of the real gains in airflow come from grinding work done to the seat and bowl area below the seat. Before he begins any grinding, Lee checks the size of the bowl, the area just under the seat, against the valve diameter. The cross section should be about the same size as the diameter of the valve. He also checks the size of the bottom of the valve seat insert against the valve diameter, this should end up at about 85 to 88 (opinions differ) percent as big as the diam-

The ring on the valve is the imprint left after Lee laps in the valve against the seat. Note that the sealing area is up near the outside edge of the valve.

This is the 30 degree cutting die used to define the upper edge of the sealing area. Next Lee will use a 60 degree angle to define the lower edge. Through this careful use of the 30 and 60 degree tools on the seat that's already cut with the 45 degree stone Lee can locate the sealing area up or down on the seat.

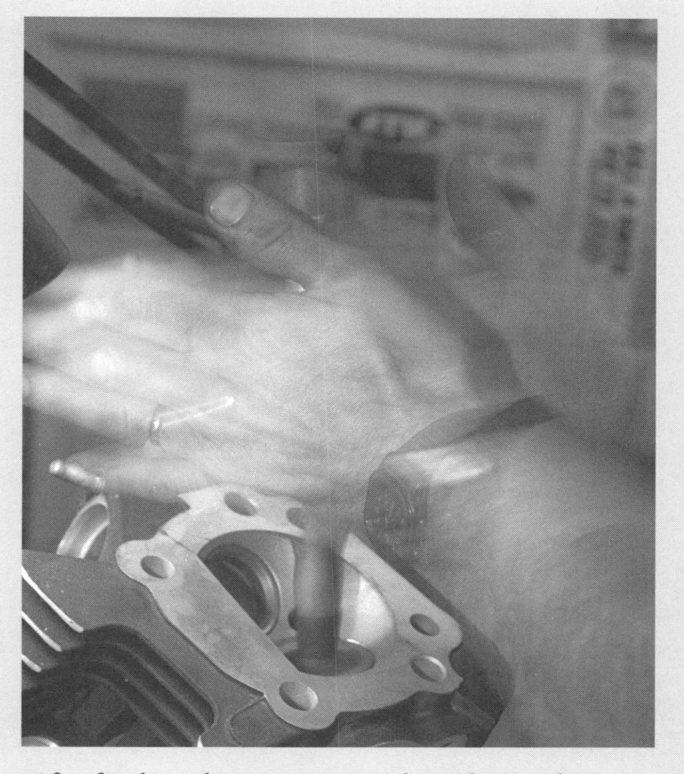

After finishing the cutting Lee applies valve grinding compound to the valve and seat and laps in the valve. This is a good way to ensure there are no high spots on the valve or seat, and also leaves the tell-tale ring on the valve that shows where the sealing surface is.

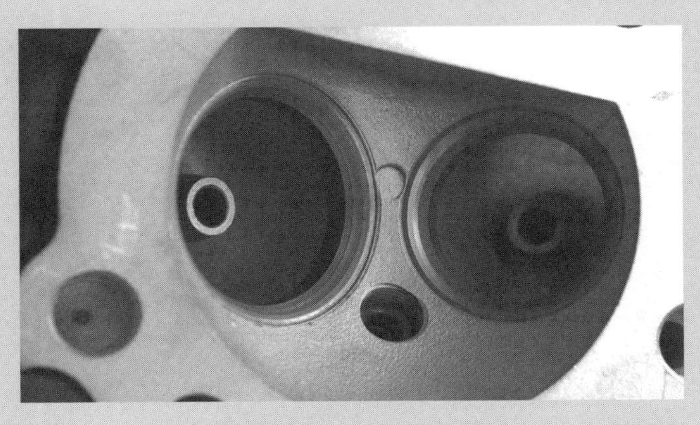

This close up shows the seat after the new valve job. Though it's hard to see, note that the actual sealing area has been moved outward on the seat. By shaping the seat material inside the sealing ring Lee will effectively optimize the seat for this valve size.

Before the grinding begins Lee measures both the bowl area and the I.D. of the seat, and compares those figures against the valve diameter.

Lee starts his grinding where the seat meets the head casting - opening up the seat I.D. and blending it into the aluminum of the head.

eter of the valve.

Lee starts the porting work by cutting down the short side radius (the "short" corner between the seat and the edge of the head) and blending the edge where it meets the seat. Then he works around the edge of the seat blending the area where the seat meets the head casting, (there is often a ledge there as the seat insert is a smaller diameter than the casting) and increasing the diameter of the seat just below the sealing area. Though it's an oversimplification, the idea is to make it easy for the air to move through the port and into the cylinder without encountering any rough edges or restrictions that inhibit flow.

Most of the real grinding is done with a coarse carbide burr, then finished with a finer carbide burr and finally a flap wheel. The flap wheel (which is used on the entire port, not just the areas where the grinding was done) smoothes the surface for better flow and also makes it easier for Lee to feel a remaining edge or low spot as he runs his finger tip down into the seat and bowl area. The remainder of the intake port is worked with the fine carbide burr for a "tooled" finish.

The Superflow 110

Before the flow bench work can start, Lee explains that , "You need to calibrate the machine periodically and you need to make the tests with the head on a cylinder that's the same size (within 1/8th inch) as the cylinder the head will be used on. I use an intake manifold and velocity stack on the head when I check intake ports."

Lee runs all his tests at 10 inches of water, as measured on the column on the left of the flow bench. Across the top we see the actual flowmeter, which reads percentage of flow. The total CFM the machine will flow varies depending on how many of the orifices are left open at the top of the machine.

Lee methodically tests each port at a series of valve openings and then with no valve at all. The readings are taken as a percentage of flow number, which is then converted to actual CFM through a series of calculations.

The worksheet Lee does on each head is extensive and contains a wealth of information for the experienced mechanic. Among the calculations that Lee does is the percentage difference between what the intake will flow and what the exhaust will flow. The exhaust should flow 75 percent or more of what the intake will flow (with it's bigger valve).

"When I look at my CFM numbers," says Lee. "I'm comparing them against other numbers that I have obtained on my flow bench. I'm careful about calibrating my machine and filling out the worksheets, so my numbers today are comparable to flow readings I took a year ago. But there are a lot of variables between machines and methods, even if they are both run at 10 inches of water, so you have to be careful when comparing the figures of different shops."

Lee reports that any good porting shop will have a flow bench to document their work, and that you should get a copy of the worksheet when you pick up the heads, so you know exactly how much they improved as a result of all that grinding.

The office at Fritz T. Wilson, complete with Super Flow bench and a computer to better use and interpret the results of the porting work.

The "ported" port, note that the seat has been fully radiused and blended into the bowl. Compare this photo to one of the early pictures.

Monometer on the left is used to keep the machine operating at 10 inches of water. The smaller manometer reads the percentage of flow which is converted to CFMs.

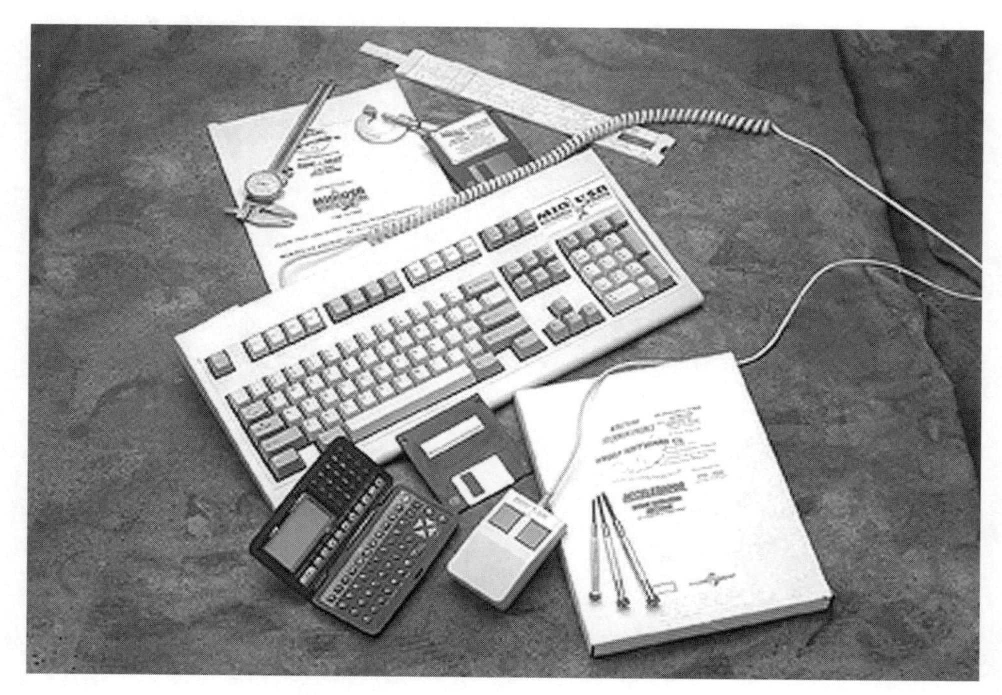

Originally designed for drag racing, the Accelerator program is today very useful for predicting the outcome in torque and H.P. of street engines. Mid-USA

Some guys buy the high-flow heads. The heads way overflow the motor. It gets back to how honest people are about how they ride and use the bike.

So if I'm building a motor from scratch I should pick the heads first?

You should check on what's available and will it work with what you're trying to build. The other thing you have to make sure of is stroke, livable street strokes, depending what you're going to do with the bike. If you're going to rebuild the thing every two years, go with a 4-3/4 inch stroke. There's a lot of controversy about stroke. Stroke will always give you more torque and more mid range. It's just because you have more volume for the same amount without having to rev it. And most guys don't ride over 6,000 rpm. If you're building for torque, you still need good airflow and a good head.

The Accelerator program gives a rule of thumb on that. It will automatically throw up a set of default values, port area, port length, bowl size and the amount of CFM that you really should be looking for. It'll give you a target.

This is a new prototype head from Craig Walters, soon to be available in the aftermarket. Note the small combustion chamber and the rectangular intake port in this new street head

Can you talk a little bit about combustion chamber shapes, good ones, bad ones.

The best chamber shapes we feel is still the stock Harley chamber. It's still a pretty damned efficient chamber.

When people describe combustion chamber shapes the term quench always comes up, can you discuss quench?

Quench is the last part of the burning process when the piston is coming up. When the piston gets really close to the flat areas of your head or whatever mating surfaces you have, it tends to excite the fuel and make it turbulent. Push it toward the desired location in the chamber. This helps complete the process of the burn quickly. The quicker you burn the fuel without detonating, the faster the energy pushes down on top of the piston instead of turning it into heat. When it burns slow you turn it into heat and then that heat goes into the dome of the piston, heats up the piston instead of pushing the piston down. Remember, heat and energy are the same thing.

When does a cylinder head need a second spark plug?

In our opinion, never. The only exception might be a Shovelhead with a big domed piston where the flame front can't get around to the other side of the piston.

The shallow combustion chambers are a better alternative to dual plugging. On evolutions, not even in our drag bikes at super-high compressions do we ever run two plugs, and we're only running 22 degrees ignition timing. You can fire ignition anywhere you want. All you really care about is at what point maximum cylinder pressure happens and at that you try to burn the fuel as fast as you can. Those are the two parables right there. Maximum cylinder pressure, which is usually 15 to 17 degrees after top dead center, and burning the fuel fast so it turns into pushing energy.

Can you discuss port design in terms of what makes a good port and the trade offs of flow versus velocity?

To define the best port is: A port that within it's size is supposed to flow a certain amount of air and does. And the valve flows the air that the port is capable of. To show somebody the absolute best port design would be a bold statement, because there are a million different ways to skin a cat on a port design. Too big - it would have to be turned to fit on the engine. Or we do have programs that can tell if your port is too big and you're not flowing, the thing's going to be a turkey. A good port design should always have a real high floor. Generally almost all of them do. A short turn. As you're looking at it, the bottom of it should be as high as possible. The top of the port should be pretty deep, to allow the fuel to turn on the top side of the port. The opening shouldn't be any bigger than the motor needs to achieve the air flow it needs.

Again, you have to either consult our program or talk to a head guy to figure out the optimum port size. These cross sectional areas have been defined through

CHART OF VALVE SIZES

	Intake	Exhaust
H-D	1.850	1.615
S&S	2.000	1.610
Merch, their most popular assembled head	1.900	1.630
STD, new assembled head	1.840	1.610
Edelbrock Performer	1.850	1.610
Edelbrock Per. rpm	1.940	1.625
Johnson	1.850	1.600
Patrick Racing	1.940	1.600
RevTech	1.940	1.610
Sputhe	1.937	1.710
Sputhe (for Sputhe 95")	2.000	1.750

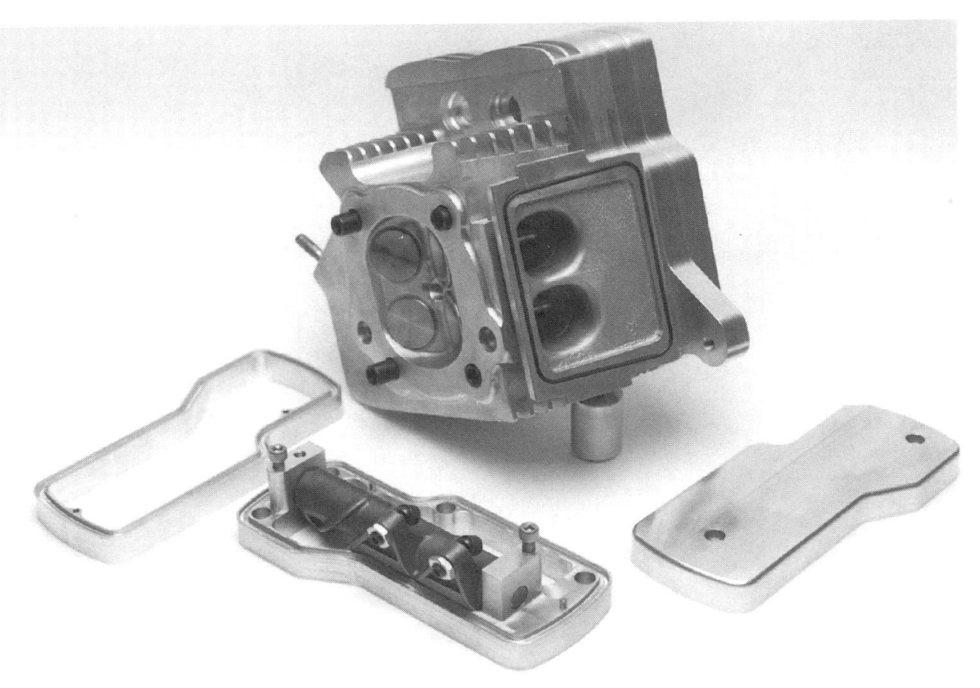

These Rivera billet 4-valve heads come in two versions. The Street model will work on bores of up to 3-13/16 while the Race version fits bores of up to 4-1/4 inches. Rivera Engineering.

years of development and hold true, whether it's a square, round, oval, rectangular—it doesn't matter.

So the different shapes and some of these double port heads that you've got are just a different way of getting at the same thing?

Yes, the same thing. Here's the thing with our four-port head. We have two inlet holes. We know we need X amount of air in a given situation. Say 280 to 300 CFM at 10 inches of water. When we look at the port area, that's a port area that's 2.5 inches wide and 2 inches tall. We have this thing called a spring pocket in there. If you go with a port that's 2 inches tall, you don't have a spring pocket anymore. Because you just bored through it. So what do you do? Well, to get the area and the velocity we put in two smaller ports of equal area, put them off to the side. Now we have room for our spring pocket. It gives us a way to run all sorts of different manifolds. We can run small inlets, big inlets. It just gives us more tunability while achieving the same air flow.

So we're back to matching the head to the engine?

You have to ask, can the engine consume the air? The only time velocity is not good is if you had a port that works good with a 1.9 inch intake valve, when you use a 2 inch intake on it and you leave the inlet the same. Now you're opening up all this valve area and you got no draw. The flow is down. What that really affects the most is when the piston is coming up from the bottom of the stroke and the intake is getting close to closing. You always look for a little bit of ram effect there. Well, there won't be any ram effect, because the velocity is not there to support the air flow. That will hurt you. It will cost you a lot of horsepower and make it feel like a slug. You have no cylinder pressure.

So if I want to put together a good combination of parts I need your program?

I wouldn't say you'd need it. It helps getting started. It also has listings from a lot of the top engine builders. There is still no substitute for good known combinations. That's the toughest thing. Unfortunately, since we do have the program I get a lot

Minneapolis Custom Cycle uses STD heads like this on many of the hot street bikes they build.

of calls about what does and doesn't work. There are certain combinations that are good and certain combinations that are bad. On certain combinations you have to compromise performance with looks. Some of the best looking exhaust pipes do not perform. That is another big scenario. I have guys running these huge, long drag pipes clear out the back that are bigger than I would use on a big-block Chevy, then when their horsepower numbers don't come up I confront them with it, and they say, "I like the looks of those and I like the sound."

If you're out racing your buddies and you're getting into that kind of scenario, then you need to find a really good, reputable speed shop. These guys deal with this all the time. Just look around and see what guys are doing. Most shops have Dyno Jets. I don't think I would deal with a performance shop if a guy didn't have access to a Dyno Jet or wasn't a pretty valid drag racing guy with experience in the gas classes.

So I should find somebody who has experience and use their expertise to get a combination that will work?

Yes. If I was going to spend $3,000 or $4,000 on something, I'd call around. A lot of guys have engine packages that they say they'll put together making a certain amount horsepower, but isn't necessarily true. Call around. Unfortunately in this industry, there's a lot of controversy on what are good and not good combinations. That's something you'll just have to sort out because advertisers have gone rabid now. I won't mention any companies, but their numbers just don't jive out. In that case, our program does help because it will tell you if the combination is going to somewhat fly. And beware. If you're going to order a set of heads and the guy doesn't want to give you flow numbers, don't buy them. Every set of heads we do gets shipped with flow sheets. Flow sheets are a disk that has the actual numbers on it. If a guy isn't willing to give you the flow numbers, don't deal with him. It's not magic. It's not secret. That tells me something suspicious about it.

As I get into more radical cams, at what point do I need to worry about valve geometry and getting the valve geometry set up when the head is assembled?

Anything over .500 inch lift, you're going to have to do some headwork. Then again, there's a point, anything over .500 lift on a stock head is questionable in terms of the added effectiveness.

You talked about exhaust. You want to talk about what does and doesn't work on the street?

Well, I could talk in a real broad sense so I don't get sued. Everybody knows that what doesn't work is drag pipes with baffles. Of course, SuperTrapp 2 into 1 pipes with a lot of disks on them work well. The Thunderheader works extremely well. Or you can put different mufflers, like Screamin' Eagles, on the stock Harley exhaust with the stock crossovers.

Craig Walters is a man with a wealth of experience designing aftermarket components for hot rod V-Twins. In order to help the end user better utilize all those components he also designed the Accelerator computer program.

Chapter Four

Camshafts And Valve Gear

When more may not be better

Camshaft Basics

Before trying to discuss how to chose the right camshaft for a given V-Twin it might be helpful to discuss the basics of camshaft operation and the terms used to describe each camshaft.

At the risk of being too simple, it might be instructive to first consider each of the four strokes and how the camshaft affects, and is affected by; the power, exhaust, intake and compression strokes. Because the camshaft runs at half the speed of the crank the

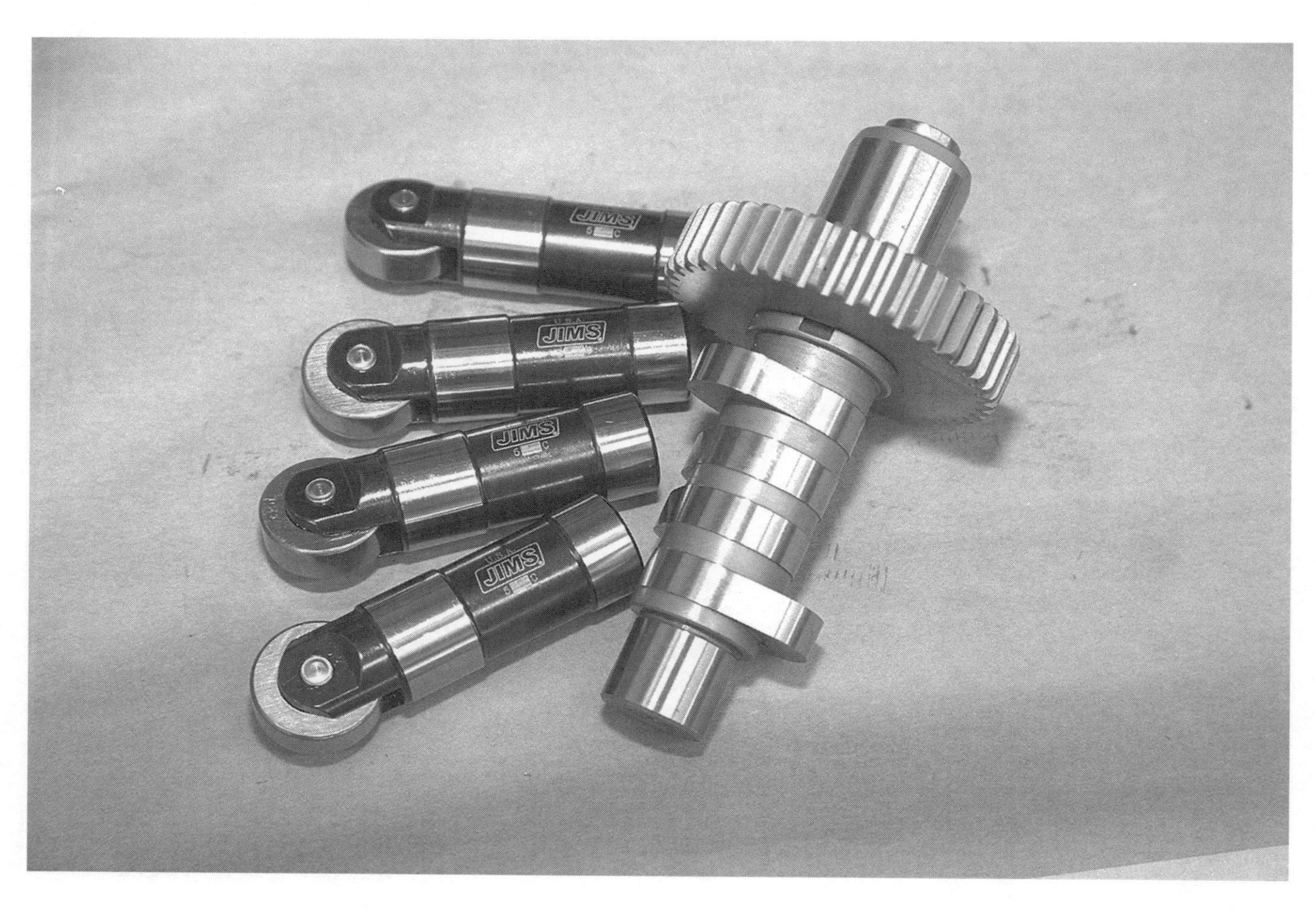

Though it sounds obvious, people forget to be sure that their hydraulic lifters are matched up to a hydraulic cam.

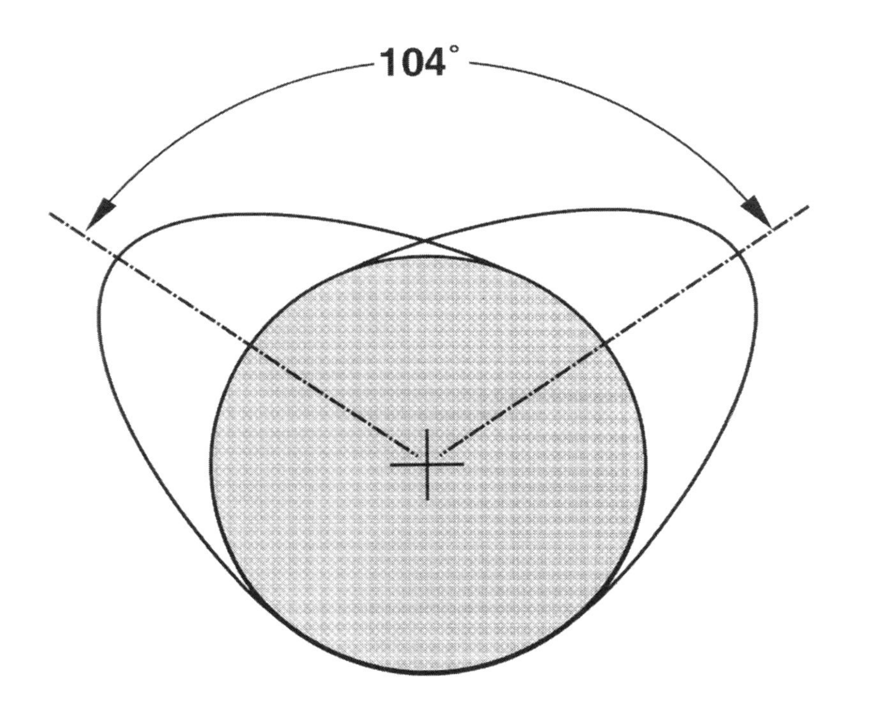

Lobe separation angle is simply the distance in degrees between the centerline of the intake and exhaust lobes. A narrow separation angle will provide more valve overlap.

camshaft will turn one complete revolution to the crankshaft's two. Remember too that most cam specifications are given in degrees of *crankshaft* rotation. Finally, the cam in a V-Twin runs backwards from engine rotation, which confuses people.

Power stroke

On the beginning of the power stroke, with the piston approaching TDC, both valves are closed. A spark causes pressure in the cylinder to build, forcing the piston down in the cylinder. As the piston nears the bottom of the cylinder the pressure, and power production, drop off. In order to get as much of the spent gas out of the cylinder as possible the exhaust valve opens before the piston actually reaches BDC. With a typical mild street camshaft the exhaust valve opens at about 60 (crankshaft degrees) before the piston hits BDC.

Exhaust stroke

The exhaust valve is open as this stroke begins, and stays open during the entire 180 degrees of the exhaust stroke. In order to pack as much fresh gas as possible into the cylinder, especially during higher rpms, the intake valve opens before the piston hits TDC. This period when both valves are open is known as the overlap period, during this time the outgoing exhaust gasses help to draw the intake charge in behind them. Typical street cams begin opening the intake valve at about 30 degrees before the piston reaches TDC during the exhaust stroke.

Intake stroke

As the intake stroke begins both valves are open. By about 25 degrees past TDC the exhaust valve has closed, ending the overlap period. The intake valve remains open during the rest of the intake stroke.

Compression stroke

Though it would seem best to close the intake valve when the piston hits BDC, the gas and air are coming into the cylinder with a cer-

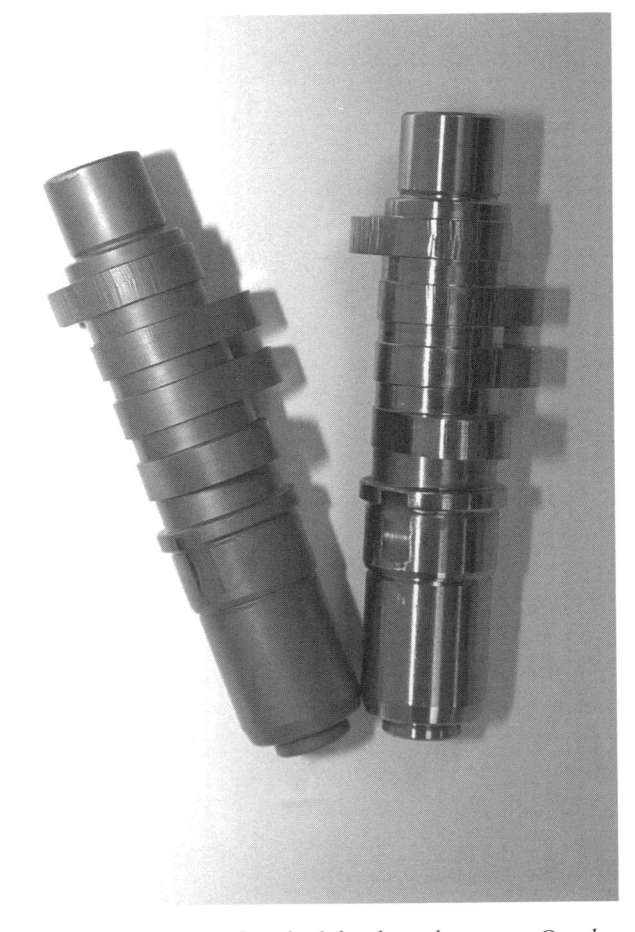

On the left, the cam after the lobes have been cut. On the right, the heat treated and polished product waiting for the gear.

tain momentum (especially at higher rpms) and the piston doesn't really start to build pressure until it has traveled part way up into the cylinder. For these reasons the intake valve stays open into the early part of the compression stroke. A typical street cam closes the intake valve at about 40 degrees after BDC.

Camshaft Terms

Valve lift

Lift is simply the amount the valve is lifted off the valve seat. The specifications are for net lift at the valve, as it is affected by the ratio of the rocker arm (V-Twin ratios for evo-style engines are typically 1.6 to 1). More lift would seem to add more power, though there are trade-offs here just like everywhere else. For example, open the intake valve too far, too quick, and it runs into the piston. Even if it doesn't run into the piston, the higher the lift (at a given duration) the faster it must move in going from the closed to the open position. Moving the valve too fast puts enormous stress on the valve train and requires stout springs, which creates even more stress. If the lobe gets too extreme the effect on the valve train is like hitting the parts with a hammer.

Valve lift is also limited by the spring. A given spring can only be compressed so far before the coils bind. As discussed elsewhere in this book, the valve lift and cam profile must be matched to the valve springs so as to avoid coil bind and keep the lifter following the camshaft even at high rpm. Most camshaft manufacturers will provide recommendations for spring pressure, and when a particular spring kit should be installed. Note:

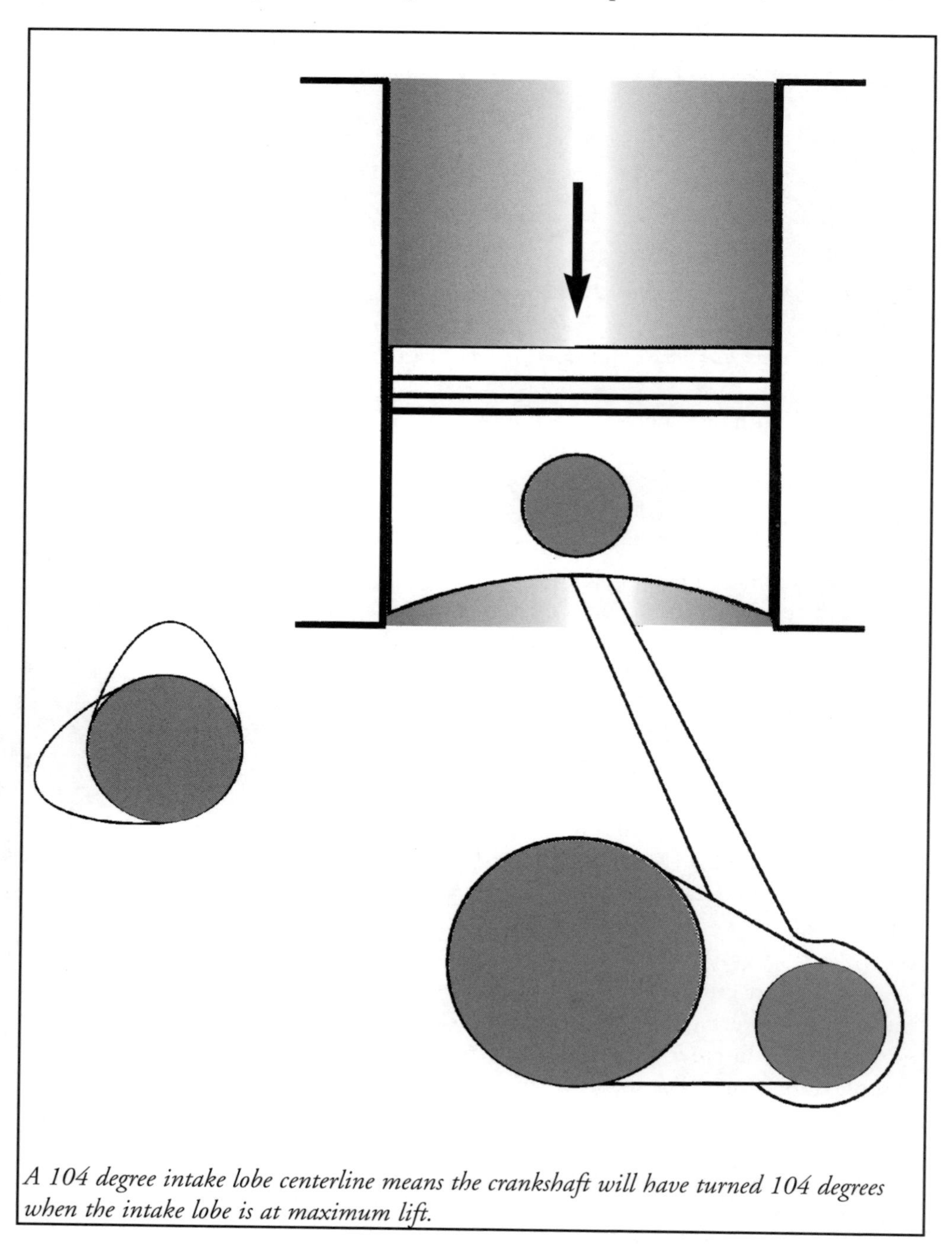

A 104 degree intake lobe centerline means the crankshaft will have turned 104 degrees when the intake lobe is at maximum lift.

With very high lifts the spring collar may hit the rocker cover.

Duration

Camshaft duration is simply the amount of time the cam holds the valve open, measured in degrees of crankshaft rotation. When comparing duration figures from one cam to another, it is important to use the correct duration specification. In order to keep everything equal most manufacturer's give a duration specification *after the lifter has achieved .053 inches of lift.* Use this specification to compare camshafts, not the advertised duration, which is usually considerably larger. More duration would seem to be a good thing for power production, except that by lengthening the duration you effectively shorten the compression stroke. Which is why you need a good match between the engine's static compression ratio and the duration.

Andrews makes a wide range of camshafts for wild or mild V-Twins with hydraulic or solid lifters. Drag Specialties

As a rule of thumb a cam with longer duration will have more power higher in the rpm range than a cam with less duration. A camshaft with less duration will tend to have better power on the bottom end and a better idle quality. Too much duration on the street will give you a bike with no bottom end - no power at lower rpm where you tend to need it most.

Overlap, as already described, is that period of time when both the intake and exhaust valves are open. As time goes on fewer and fewer manufacturers provide any overlap specification.

Lobe Separation Angle

This is the newer specification that is often provided instead of overlap. It is the distance in *cam degrees* between the centerline of the intake and exhaust lobe. A 104 degree lobe separation angle means the centerline of the intake lobe is separated from the centerline of the exhaust lobe by 104 degrees. Similar to overlap, the separation angle is more comprehensive. Not only does it include overlap information (a narrow angle means more overlap) it provides more valve timing information than the simple overlap specification alone. In general, a cam with more overlap and a smaller separation angle will tend to have a power band that is peakier and occurs at a higher

These roller-tip rocker arms from Crane come with their own shaft and needle bearings to further reduce friction in the valve train. Drag Specialties

rpm than a cam with less overlap and a larger lobe separation angle.

Intake lobe centerline.

This is a timing specification and gives the position of the piston when the intake valve is at its maximum opening. If the intake lobe centerline is 104 degrees the crankshaft will have turned 104 degrees past TDC at the point when the intake valve has maximum lift. Measured in crankshaft degrees, intake lobe centerline can be compared with the lobe separation angle to indicate the amount of advance the camshaft has relative to the crankshaft.

Mid-USA markets their own line of camshafts, from "bolt in" to street and strip, under the Power House name. Mid-USA

Fully understanding how intake lobe centerline affects engine performance requires a short explanation. At TDC, or 0 crankshaft degrees, the cam has both lobes pointed up, this is the middle (or roughly the middle) of the overlap period and also places the cam in the approximate middle of the lobe separation angle. If the same cam with an intake lobe centerline of 104 degrees had a lobe separation angle of 104 degrees it would be said to be 0 degrees advanced.

RevTech makes these adjustable pushrods from 7/16 inch O.D. aluminum tubing. Pushrods commonly come in 3 lengths: the longest goes to the front exhaust, the next longest to the rear exhaust and the other two for the intakes. Custom Chrome

In other words, starting at TDC, the piston moves down through 104 degrees of crankshaft rotation, or 52 degrees of cam rotation to put the intake valve at maximum lift. Because 52 is half of the lobe separation angle, the cam was in the middle of the overlap period (half the lobe separation angle) when the piston was at TDC. If the same cam with the lobe separation angle of 104 degrees had an intake lobe centerline of 100 degrees, then the cam is said to be 4 degrees advanced - this cam would reach maximum intake lift after 100 degrees of crankshaft rotation, or four degrees sooner.

Putting it all together

Designing a cam is very much a matter of matching the lobe shapes, lift, duration and timing to a particular set of operating conditions. Timing is always critical on a V-Twin cam to ensure that the two valves don't run into each other. To produce power air and fuel must be encouraged to enter the cylinder. Once burned those fumes must likewise be encouraged to exit the cylinder so another fresh charge can enter. Considering only the over-simplified situation here long duration and high lift would seem to produce more power, until you realize that for everything you gain, you likewise give something up.

In particular, an exhaust valve that is open too far into the intake stroke, for a lengthy overlap period, might work well on a high rpm drag race motor. At low rpm however, the engine has low vacuum (the exhaust valve is open for so much of the intake stroke that it acts like a giant vacuum leak), runs rough, and will probably spit exhaust pressure into the carburetor and/or push raw gas right out into the exhaust pipe.

This 3-piece spring kit will accept cams all the way to .600 inches of lift with higher spring pressure than standard springs. 10 degree keepers, rather than the normal 7 degree, give greater security. Custom Chrome

The cam that's right for your bike will depend on a whole list of variables, including the carb and pipes, the compression ratio, the shape of the ports, valve size and the type of riding you do. Getting that perfect cam means asking the right questions and being honest about the type of power you want and the kind of riding you do. The engineers at S&S add the comment that engine displacement is a very important consideration, because bigger engines can better utilize cams with longer duration and higher lift, and that weight and gearing should also be considered when trying to choose the right camshaft. Don't make the mistake of mis-matching a camshaft meant for solid lifters with a set of hydraulic lifters.

You don't need solid lifters to go fast - solids can rev to 6000 rpm and more with no valve float. This hydraulic lifter from Jim's is designed to fit all Evo V-Twins. Drag Specialties

Which Camshaft Is Right For You

The new Crane Cams Motorcycle catalog has a long

Camshafts must be made from very high quality steel with little or no impurities. These camshaft blanks wait for further grinding at the Andrews plant.

list of things you should consider when choosing a camshaft for your particular bike. The comments (which echo those of other builders interviewed in the book) list the important criteria as: riding style, weight of bike, operating range, engine displacement and compression ratio, and engine modifications and accessories.

They go on to explain that when it comes to camshafts, " the 'bigger is better' belief... is usually far from the truth." and that, "The camshaft provides an 'rpm power band' that is approximately 3,000 rpm wide. This rpm power band can be produced in either the lower range (idle to 3500 rpm), in the mid range (2000 to 5000) or in the high range (3500 to 6500)."

The engineers at Crane advise that achieving a particular performance goal means you must produce the right amount of power at the correct rpm range. If your bike is heavy they advise using a cam with less duration as an aid to get that weight moving.

Compression pressure is often affected by the camshaft choice. In particular, cams with more duration result in decreased cylinder pressure, in which case you would want to offset the loss by designing an engine with a higher compression ratio. The catalog explains that, "This is why the descriptions of the camshafts may note that an increase of compression ratio is recommended with a particular design."

Before closing the section on camshaft choice, the catalog explains that bigger engines (more cubic inches) work better with camshafts with increased duration and lift. Their final piece of advice

This is a camshaft being ground on a standard cam grinding machine at Andrews. This type of machine follows the contours of a "master" cam as it grinds.

should perhaps be set in boldface type. "It is important to put correct component parts on the engine that will work together and enhance the desired performance."

CNC style Landis cam grinding machines have replaced most earlier style machines at Andrews. These require no master, and machine lobes with great accuracy.

VALVE TRAIN COMPONENTS

The more radical the valve train geometry, the greater the need for roller tip rocker arms. The roller tip minimizes wear on the valve stem tip and side loads on the valve stem. Full needle-bearing rocker arms further reduce friction in the valve train. Though some competition arms come in non-stock rocker arm ratio, the selection that follows comes in the standard, 1.6:1 ratio. As lifts get more extreme there may also be a need for rocker arms with corrected geometry that puts the rocker arm half way through its working cycle when the valve is halfway between open and closed.

Baisley

This company takes the stock rocker arms you supply and adds roller tips. While they are attaching the new roller tips any necessary corrections to the geometry can be performed at the same time.

Crane

Well known as a manufacturer of camshafts and related components, Crane makes two styles of roller-tip rocker arms. The "bushing" roller tip rocker arms fit 1984 and later V-Twins and use the standard rocker shaft, which passes through the bushing in the arm itself. Crane also makes needle bearing roller tip rocker arms. These low-friction rocker arms with roller tips are supported by needle bearings instead of the standard bushing and come with their own hard-

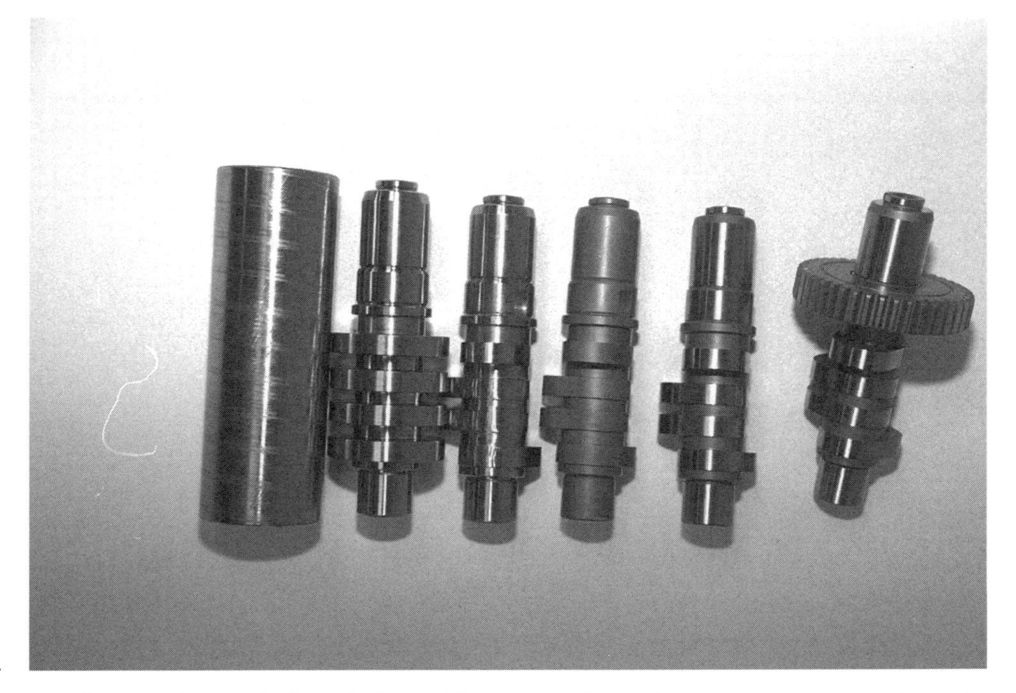

Evolution of a camshaft at Andrews. Here we see the various stages a cam goes through from a bearing-quality steel blank to a finished camshaft complete with press-on gear.

ened shaft. No machining is required for installation.

Rivera Engineering

Rivera makes a set of adjustable rocker arms designed to improve the geometry of the valve train. Each rocker is designed for a specific valve, and is further adjustable three degrees in either direction. They also manufacture a gauge to make it easy to check the geometry of each valve.

S&S

Bushing-style roller-tip rocker arms are available from S&S. These rocker arms use stock rocker arm shafts and fit 1984 and later V-Twins.

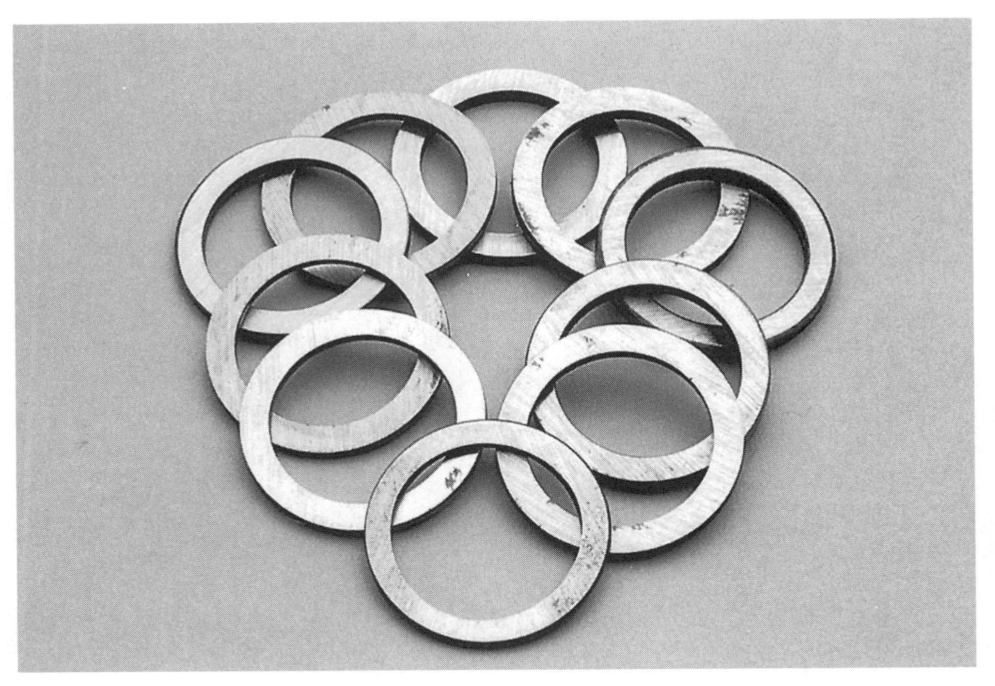

Thrust washers of various thicknesses are available to help you correctly set the end play of the camshaft. Custom Chrome

Pushrods

There are too many pushrods on the market to list them all here. Some pushrods have an enormous amount of adjustment and are meant to facilitate installation of a new cam without removing the rocker boxes. These are not what you need if you are building a motor from scratch.

Adjustable pushrods come in chrome-moly and aluminum with 1/4 inch or 5/16 inch adjusters. You don't want pushrod flex, which robs the valve of its travel, but you likewise don't want a super heavy duty pushrod that weighs five pounds. You simply need a good sturdy, adjustable, pushrod, which most of them are. The final choice will depend on personal opinion and availability.

Roller tip rocker arms are often used on high performance V-Twins to reduce friction in the valve train.

Interview: John Andrews, Andrews Products

John Andrews is the founder and the driving force behind Andrews Products, the well-known manufacturer of gears and camshafts for V-Twins.

John, how'd you get into grinding and designing cams?

I started making gears in 1972. Later it occurred to me

that the manufacturing technology for making cams and gears isn't completely different. The same machines make the blank. Same machines do the finish, bearing, grinding. In the middle there's a group of machines that put gear teeth on one group of parts and cam lobes on another group of parts. We got into cam grinding by buying some prototype parts and then we bought some regrinding machines and adapted them to short run production and just kept upgrading the equipment. We started with cams in 1974.

So you got into the market through the gear end?

Right. The old four-speed Shovelhead transmissions had these truck transmission ratios so we made some first gears and everybody thought they were a revelation, but all it amounted to was taking two teeth off the counter shaft gear and putting them on the cluster gear. They made big twin engines feel like a Sportster. It shifted up through the gears real nice which is apparently what everybody wanted to do anyway.

At that time you talked to the factory about making this ratio?

Yes. They didn't want any part of it. After the dealers liked it they decided it was a good idea, but it still took five years for them to get it done. I made gears in 1972 that took until 1978 for the factory to get them on a Superglide. Dealers were taking brand new bikes out of the shipping cartons and taking transmissions apart and switching first gears, because once the guys got used to our first gear they didn't want them the other way. It's been a good market for us.

How did you sell the gears, how did you get started?

I used to travel around the Midwest for my old job, as a consulting engineer. I would take a gear with me in my briefcase. I remember stopping in at a dealer in Tulsa, Oklahoma. They had the tin ceilings and the glass cases, the tile floors - this older man and his wife ran the place, and I set the gears on the counter and the guy walks over and says, "Let's see. What have you got here, junior?" I said, "First gears." He said, "What do they do?" and I said, "They straighten out the first gear ratio on a Superglide." He says, "What do you want for these?" I had two sets of them. I told him $65 a set. He said, "You can leave them." He turns around and told his wife to write me a check for $130.

You were not trained as a machinist?

I had some background as a machinist, but I was trained as a mechanical engineer. I went to school when they still thought it was worthwhile to teach us how to use a lathe and a milling machine. It stopped in about 1970. I know engineers who went to school after 1970 who wouldn't know a milling machine if it fell on them.

A lot of people don't understand the timing and duration and catalogue numbers. Could you walk through that real briefly?

These numbers represent where the intake and exhaust valves open and close. They are really sort of an abbreviated statistic for what people think of as a cam. A lot of people call here and want to compare our numbers with other people's numbers. Because of the way the numbers are generated that is hard to do - numbers between manufacturers are somewhat comparable but not directly comparable. They're not comparable in all aspects.

Looking at a cam where it opens and closes at a checking height of .0050/.0053 inch doesn't tell you what it does. Below that it doesn't tell you what the shape of the tip is and it doesn't tell you how well the part was made. It's just a very rough statistic. The auto-

Gary Grimes from Minneapolis recommends adjusting the valves by: First finding the low spot on the cam, then adjust that pushrod to zero lash, next lengthen the pushrod according to the manufacturer's instructions (this will be different depending on the pitch of the threads).

motive industry has been doing this for years, so that's how it came to be. The most important number is the point at which the intake valve closes because that's where the piston starts pumping compression pressure. If that isn't right nothing else in the cam is going to make any difference.

Do you want to address the timing that works best on a V-Twin?

Harley-Davidson engines work best anywhere from 102 to 108 degrees depending on the size of the engine and the specific application. With Sportsters it's a little less. I know car engines today still run anywhere from 112 to 118 degrees. Chevy race car motors - those get down to 106/108 degrees. It may have something to do with the inefficiency of the engine. The more inefficient the engine, perhaps the angles can be closer up to a point.

Some camshafts come with recommended spring kits as a way to ensure there is enough spring pressure (you should still check both pressure and coil bind).

You touched on compression. Do you want to speak to the relationship between compression and cam shaft design?

If you're going to take a bike that's 8.5:1 compression ratio, which is what some of the stock Harley-Davidson's are, if you put too much cam in it you aren't going to get enough compression to make it run right at low speeds. If you don't put enough cam in it you'll get too much compression, and it'll be maybe peppy up to 3000 or 4000 rpm. After that it won't have anything. So it's somewhat important that the cam duration and timing be reasonably matched up to the engine/compression ratio. What's acceptable on a drag strip probably won't be very much fun on a street bike.

Is it a mistake when people use drag strip cam shafts and

In addition to camshafts, Crane makes a variety of valve spring kits. Look to the cam manufacturer or someone familiar with head work for a spring recommendation. Drag Specialties.

engine technology on the street?

Yes. There is this feeling that if some is good, more is better, and too much is just enough. That's not true. We have a lot of guys that over-cam engines and then tell us the engine feels real good at 5,000 rpm, but at 2,000 rpm their friends with a mild cam beat them. You can feel the difference, but until people see it side by side they don't believe that it's true.

Weight is an important consideration in figuring out which cam to use?

Yes. If you're pulling a lot of weight you're better off with a mild cam. It's going to have a broader torque curve and the power will come out of the lower rpm range where you really want it. If you have a lightweight bike on a drag strip it doesn't matter if it doesn't start pulling hard until it gets to 4,500 rpm because it won't ever be below that anyway.

Black Nitrided valves provide better resistance to wear and corrosion than stainless or chrome.

When I buy a camshaft, how am I sure that I get the right springs and the right spring pressure?

We try to design cams so that they don't require over springing. If you set up the springs, set them so that the cam runs maybe up to 6,500 rpm without separation, to get above that you're going to have to get some pretty brutal springs. Those are hard to install. Sometimes they have a short life and excess spring force is just going to waste horsepower. You're just going to be sitting there compressing springs and all that spring compression basically ends up as wasted heat. What you need is just enough spring and no more.

If I'm buying a cam I should look to the cam manufacturer for

Crane makes two series of camshafts: The Fireball has a standard pressed on gear while the Hi-Roller uses a keyed gear which makes it possible to adjust the cam timing 4 degrees either way from zero.

This steel breather valve features adjustable timing to compensate for larger displacement engines. Comes with an assortment of thrust washers. Custom Chrome.

a recommendation?

Yes. The cam manufacturer should know. On some of our cams the lifts are the same as a stock cam, but the duration's are longer. It makes it fairly easy to use stock springs because what you're doing is telling the spring you're only going to lift to the same point but now you've got more time to set the valve down, which actually requires less acceleration and less spring force than a stock cam.

If I'm putting together a new motor, do I always need to check for coil bind and the seated pressure of the spring?

If you're putting a cam in that requires spring changes or new valves and new seats you should always check it. If people are just going to take a relatively new bike and put a cam in it, as long as the timing and the lift isn't something wildly different than stock, it should fit. When we design cams we always try to make sure that the piston-to-valve and valve-to-valve clearance numbers are not going to put people in trouble with a stock engine.

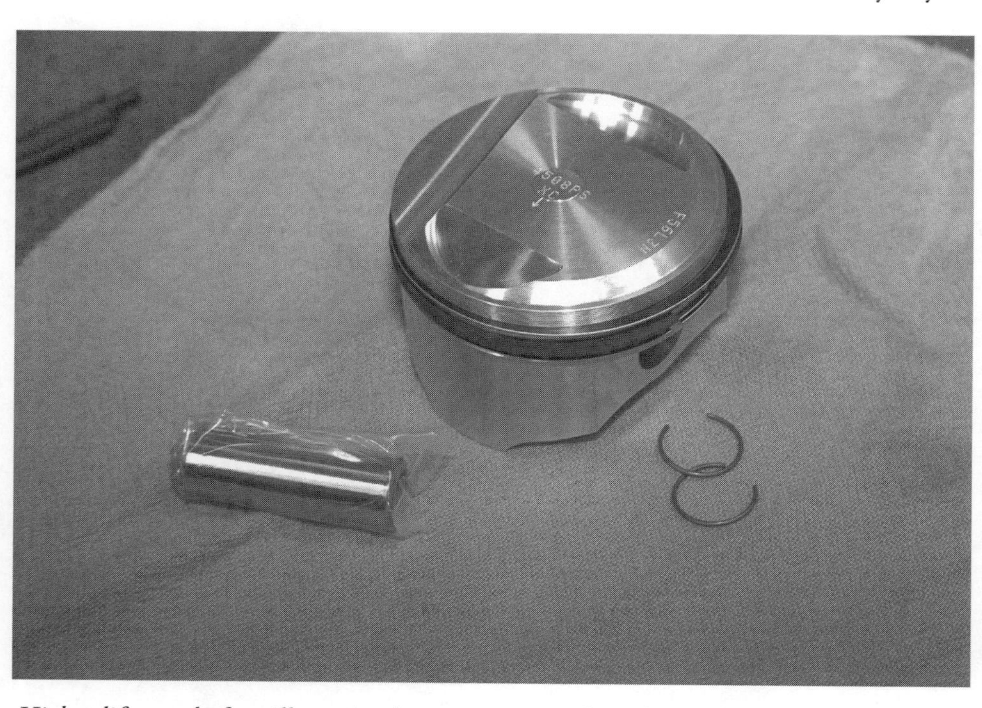

Higher lift camshafts will require deeper cutouts in the valves. Most builders recommend you mock up the entire engine before final assembly with clay on the cutout to ensure enough clearance.

If I'm doing a kit thing - maybe I buy some RevTech heads that are already assembled, do I still need to check for coil bind and the spring pressure?

It probably doesn't hurt. Once the parts are disassembled it's pretty easy to check them. It's a lot easier to check them than it is to do them twice. A lot less expensive too.

There seems to be a fair amount of confusion about gear fitment on a new unit assembled from scratch. What would your recommendation be in terms of cam gear to pinion fitment?

This is a question we get into everyday on the phone. My feeling is that in the absence of a stock engine that you already took apart, where it was running quietly, you have no choice but to take two parts, a pinion and a gear, put them together in the engine and feel the fit. It's possible to put indicators on it which some people may not have, but if you put the engine together and you notice it's real tight that's fairly obvious, then you'll have to go to a smaller pinion or smaller cam gear. Somehow the fit on the gears is going to have to be brought to zero backlash when the engine is cool.

I've had mechanics tell me that they like to reach in through the lifter bores during the assembly, and there should be just a little drag between the gears as you slide the cam back and forth the long way.

If you're sliding the cam in and out of the engine and you can feel a slight drag between the gears, that's probably a pretty good fit. You can also reach in and rotate the cam forward and backward. If you can feel backlash with your fingertips it's too much. Ideally the backlash should be zero. If the engine whines after it's started the gears are too tight. That can cause damage. If the fit is too loose it can get into some rattling. It's annoying, but it won't hurt anything. It usually occurs only at low speeds.

If my fitment isn't right, how many different cam gear sizes are there? How many different pinion sizes are there? It seem to be a little confusing in terms of how to measure them, and there's two different pin sizes to be used during the measuring?

Harley-Davidson makes seven pinions and seven gears. We make three gear sizes. There is also a confusing point, and I'm not sure how this came to be. The aftermarket is selling pinions and cams that are "matched up." I don't see how they can do this with any degree of accuracy. The only way you can tell if the pinion and gear are correct is if you already knew what the center to center distance is in that engine, and since they don't know what engine it's going into, they have no way of knowing that. So how are they are going to hand you two pieces and tell you that the match is correct?

My emphasis is on hot street motors. Are hydraulic lifters good enough for most of those applications? Is there any point in going to solids?

Yes to the first question. No I don't think there is a point to using solids on the street. Hydraulic lifters are good to at least 6,000 rpm and maybe more. American cars use the same technology, Chevy V8s go way past 6,000 rpm. The problem with mechanical lifters on a air cooled engine is that the engine expands and the lifters don't expand to match. The net effect is you're going to lose lift and duration.

I hears some horror stories about cams that lose lobes after a relatively short life span. Is this typically a problem of too aggressive a cam, or not enough oil changes, or what?

I would say that it's probably more related to material and processing at the time of cam manufacture, or people have set up an engine with insufficient spring travel and the springs are stacking solid. The valve springs are stacking solid maybe .010 inch before the cam actually reaches maximum lift. It won't cause rocker arm failure and it won't cause parts in the engine to break up, but over a short time it will cause the tip of

This new cam is ready for installation (be sure to use recommended pre-lube) with the thrust washer (in the correct thickness) and thrust plate in place.

the cam to fatigue and show pitting and failure, just from being overloaded in that one spot.

Sometimes if the cams aren't ground with a true smooth acceleration curve that can affect them too. Over a short time they'll show pitting. Any pitting on the surface of a ground metal, bearing, or pin, is an indication that the surface has failed.

People need to make sure they have .050 or .060 inches of clearance between maximum valve travel and coil bind. That way the springs won't fatigue from running right at coil bind.

Can you talk about ramps, some open faster than others, are some too aggressive?

We can design cams that have acceleration curves that would put push rod systems into an overloaded condition. On some engines that have direct overhead valves you can get away with a whole lot more acceleration than you can on a Harley-Davidson, mostly because those engines don't have the long push rods. Everything is just too flexible with a Harley-Davidson engine. So within a range, we know what works and what can be used without risking failure of the parts or too much abrupt change on the acceleration curve.

All of our cams have some sort of clearance ramps at the bottom. This is the flank of the cam lobe where it meets the base circle. It's just a little transition. Kind of a flair-out where the parts are slowed down or an opening part gets accelerated, kind of gradually increasing level so that you're not just shocking them every time the cam roller follower comes around. On the closing side, if you set the valves down too hard, within a short time you'll see the valve pound the seat out. We've seen engines that have had that happen to them. Some cam grinders think that if they just put a smooth curve between a high and low point that's all they need, but that won't work. Some of our cams have fairly sophisticated mathematics designed to slow down the acceleration curve as the closing side of the ramp meets the base circle so that you're not just dropping the valve on the seat.

Do you favor aluminum over chrome-moly in terms of push-rod material, and is aluminum advantageous because it expands at the same rate as the cylinders?

In terms of strength aluminum is OK but chrome-moly is probably a much more rigid push rod. And with the hydraulic lifters, it doesn't matter which one expands more.

What are the common mistakes people make when they chose a cam?

The most common one is probably looking for too much cam. We have guys call up with a stock motor and they want to put cams in that are 20 or 30 degrees longer than what they ought to have in there. And if you try to tell them they'd be better off with a cam that gives them a nice broad torque curve they don't want to hear it. They just want a big cam, and then they find out their friends can beat them with a mild cam and they're back on the phone.

The one big mistake they make is not knowing what rpm range they live in. Because they all think they're bouncing off the rev limiter when they ride. The guys with Softails don't have tachometers,

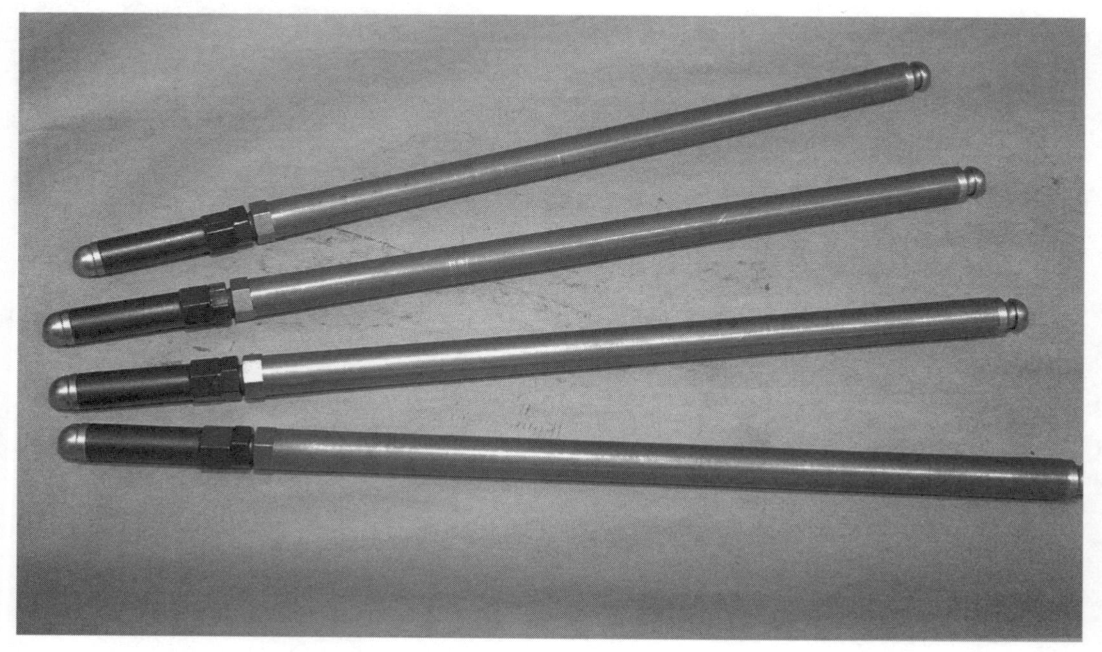

Adjustable pushrods come in chrome moly and aluminum. Which is better depends on who you talk to (for mild cams it probably isn't critical).

and if you look at mph they're shifting at they're really shifting at 3,200 rpm, so a cam that operates at 5,000 doesn't do them any good.

If you take two bikes that were identical and you put one of our mild cams in one and a big cam in the other one, if you wound up the engines in third or fourth gear or maybe 4,000 rpm and then took off, the big cam might win. But if you're coming out of a corner with a passenger and both people just turn the throttle on in third or fourth gear, the mild cam would pull away every time. Some guys understand this, but sometimes it's a hard concept to sell.

People should look at the installation and checking procedures outlined in your catalogue?

Yes. We try to outline the right procedures with at least a superficial discussion so that people have some awareness of these things. The other thing is, and I think people realize now it doesn't work, there were people at one time selling interference-wound springs. This absolutely won't work. Interference-wound springs, the idea of the inner spring being forced into the outer spring so there was a tight fit...over a period of time it just fatigues both springs and they break.

Do you want to talk briefly about what your cam shafts are made out of?

We use 8620 steel which is purchased as bearing or aircraft quality specification steel, and that means there's a certain limit to how much impurities there are in the material. No steel you buy is going to be absolutely molecularly clean, it's going to have some trace elements and some non-metallic particles in it. There's a specification for how much non-metallic stuff can be in the steel without exceeding the specification for this material, and we only buy that material.

Some people have tried to use material with leaded alloys, which makes it easier to machine. But Harley-Davidson cams, even thought they're small, are heavily loaded as far as stress in pounds per square inch. They're much more highly loaded than an automotive cam because of the size and the small width that the lifter rides on. What may work on an automotive engine in the way of lower quality material won't work on a Harley-Davidson cam.

Yours are heat treated?

Yes. They are heat treated and treated in a manner that prevents certain kinds of metallurgical conditions from occurring, like retained austenite, and after that they're finish ground on the bearing and the cam surfaces. I would match our finished quality against anybody in the industry, in the world really. The material we're using is pretty much accepted as the best cam material going.

John Andrews has been making camshafts for V-Twins since 1974. Today John is still making camshafts, and selling his own computer program to aid other engineers with camshaft design.

Chapter Five

Carburetors

The mixing bowl

What Is A Carburetor

A good carburetor is one that will supply a regulated quantity of air and fuel to the engine under all operating conditions. This means that no matter how hot or cold the engine is or how quickly you whack open the throttle, the engine will receive the correct amount of fuel for optimum operation. A tall order indeed.

At the heart of nearly any carburetor is a venturi, or a restriction of some kind in the carburetor throat. As Mr. Bernoulli discovered more than 200 years ago, when you force air through a restriction in a pipe the

For competition work some builders like to use the "D" carb from S&S (seen here with a Thunder Jet at Minneapolis Custom Cycle).

speed of the air increases as it moves through the restriction. As the velocity goes up the pressure goes down, think of the fast moving air as being "stretched" so it has less pressure. Now if we introduce gasoline at this low pressure point in the venturi, subject to atmospheric pressure at the other end, the gas will be "pushed" out into the airstream where it can atomize and mix with the air on its way to the cylinders.

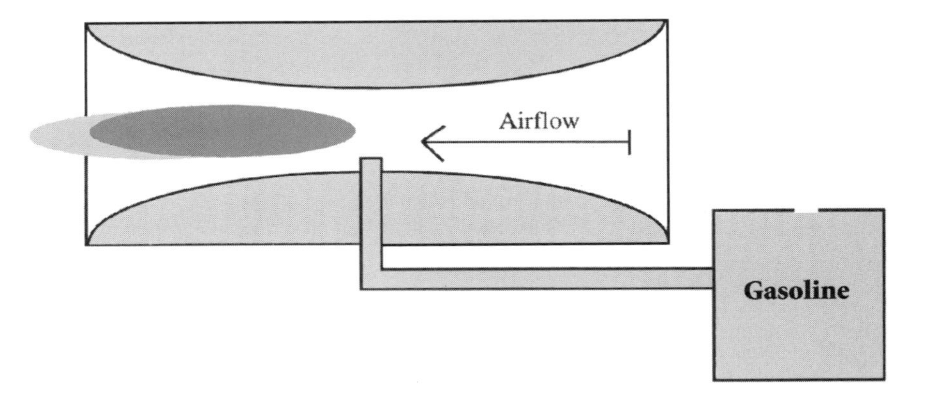

At the heart of every carburetor is a venturi, a restriction in a pipe. Air pressure within the venturi is reduced so the gas, which is under atmospheric pressure in the float bowl, flows to the venturi where it mixes and atomizes with air in the carburetor throat.

The simple carburetor described above might work on a constant speed engine, one that was always running and never changed speed. In the real world we need a means of controlling the flow of air through the carburetor and some additional fuel circuits. The extra circuits are handy for those occasions when there isn't enough air moving through the carb to create the vacuum needed to draw fuel from the venturi.

A good carburetor is designed to work in three "conditions," often described as: idle, low speed and high speed. Though many carburetors have these three basic circuits, and more, with air-bleeds and other provisions to ease the transition from one circuit to another, some are "seamless" examples that posses only one tapered needle that moves farther and farther out of the main jet in response to either throttle position or engine vacuum.

Even with a provision for controlling the airflow and various speeds real-world carburetors have to deal with situations like cold starts and sudden acceleration. A cold engine needs an extra-rich mixture for example, because gas doesn't

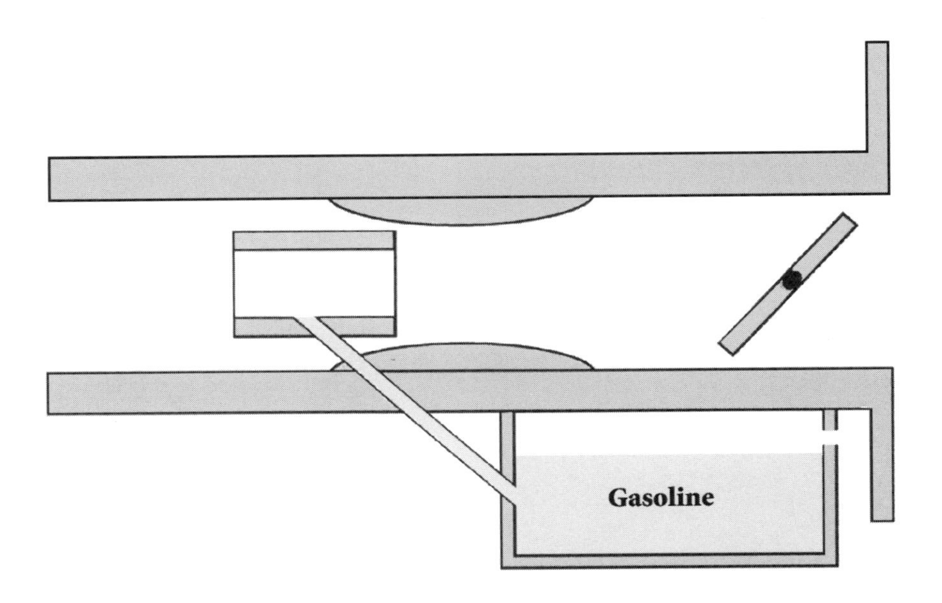

This is a simplified look at a fixed venturi, or butterfly, carburetor. Gas flows from the float bowl to the venturi, how much air passes down the carburetor throat is controlled by the butterfly valve.

like to atomize with cold air. Sudden acceleration, on the other hand, means the amount of air passing down the throat of the carburetor increase instantly, while the heavier fuel takes considerably longer to catch up.

So we add a choke or enrichment circuit for cold starts and an accelerator pump, though some carbs don't need one, to squirt a little extra gas down the carb throat when the guy in the Buick suddenly comes up along side your bike.

When trying to describe carburetors it seems there are as many exceptions as there are rules. So rather than continue describing a theoretical carburetor it might be easier to jump right in with descriptions of the various types of carburetors and follow that up with a brief explanation of each of the most popular carburetors currently on the market.

Butterfly Carburetors

A fixed venturi carburetor, also known as a butterfly carburetor, has a fixed restriction in the throat. A "butterfly" valve is used to control the amount of air flowing through the carburetor throat. Fuel for high speed operation is usually introduced at the venturi while fuel for idle and low-speed operation is often introduced into the throat closer to the butterfly valve.

Supporters of this design cite the fact that butterfly carburetors have been used on everything from Model A Fords to Harley-Davidson motorcycles. In the aftermarket, the S&S, RevTech, Bendix, Screamin' Eagle and several others are all butterfly designs.

Constant Velocity And Slide Carburetors

Some carburetors vary the size of the venturi in the carburetor throat according to throttle position or engine load. These variable-venturi designs come in two basic models (more later). Usually the slide or variable restriction is connected to a tapered needle that passes through the main jet. In this way increases in venturi size are tied directly to increases in fuel.

While fixed-venturi designs see changes in air speed through the carb throat as the throttle is opened or closed, variable-venturi designs keep airspeed nearly constant through the carburetor throat.

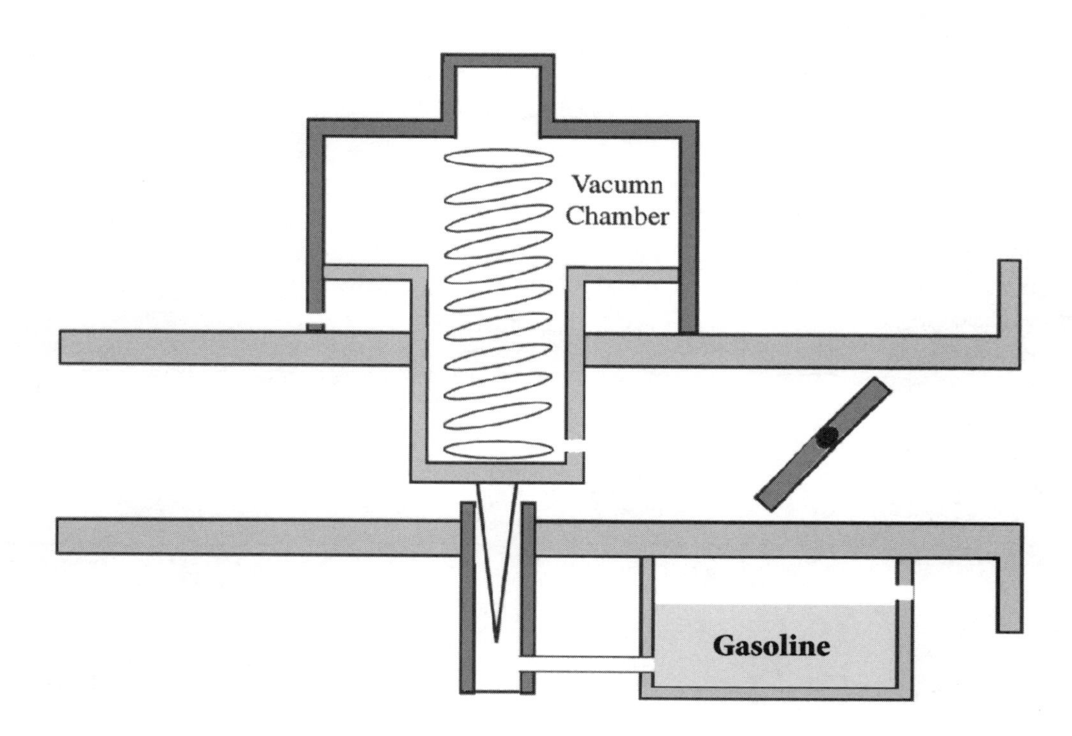

A Constant Velocity carburetor uses a vacuum controlled slide and a standard throttle valve to control air moving down the carburetor throat. At idle a spring holds the slide closed, as the throttle is opened and the engine gains rpm, increased engine vacuum pulls the slide up against spring pressure. The slide moving up pulls the tapered needle farther out of the main jet. Because the slide moves with engine load and speed the velocity of air in the carburetor throat stays constant, hence the name.

Mikuni makes slide-type carburetors in a variety of sizes for V-Twin applications. Older kits and V-Twins used a single cable though now nearly all carbs and kits are designed for dual cable operation.

There are two styles of variable-venturi carburetor: the constant velocity design, which we often call a CV carburetor, and the straight variable-venturi design, sometimes called a slide carburetor.

In the constant-velocity design the throttle is connected to a conventional butterfly valve. Upstream from the butterfly valve is the variable restriction in the carburetor throat. This restriction is held in the closed position by a spring and opens according to vacuum within the carb throat. More vacuum causes the piston to open farther, increasing the size of the venturi. At idle for example, both the butterfly and the venturi are closed. As the throttle is opened more vacuum is applied to the slide piston, the piston moves up until equilibrium is achieved between the spring pushing down and the vacuum pulling up. As the slide or piston moves up the tapered needle is pulled out of the jet, effectively increasing the size of the jet. The most common CV carburetors in the aftermarket are the SU and the billet carburetor from Carl's. The Keihin carb used on factory bikes since the mid-1980s is also a CV design.

The non-CV carbs with variable venturi are usually known as slide or smooth-bore carburetors. These designs eliminate the butterfly altogether and connect the throttle cable to the slide. The slide is connected to a tapered needle that passes through the main jet. As you open the throttle the slide opens the venturi allowing more air through the carburetor throat. At the same time the tapered needle is raised in the jet, effectively increasing the size of the jet and adding more fuel to the increased air flow. Slide type carburetors include the Mikuni, the QwikSilver 2, and the Dell'Orto "pumper."

Fans of the variable-venturi carburetors, both CV and slide type, point out the simplicity of the design. A design that eliminates most of the

The new Typhoon from Carl's Speed Shop is a CV design carved from billet aluminum and designed to work on a wide range of V-Twin engines.

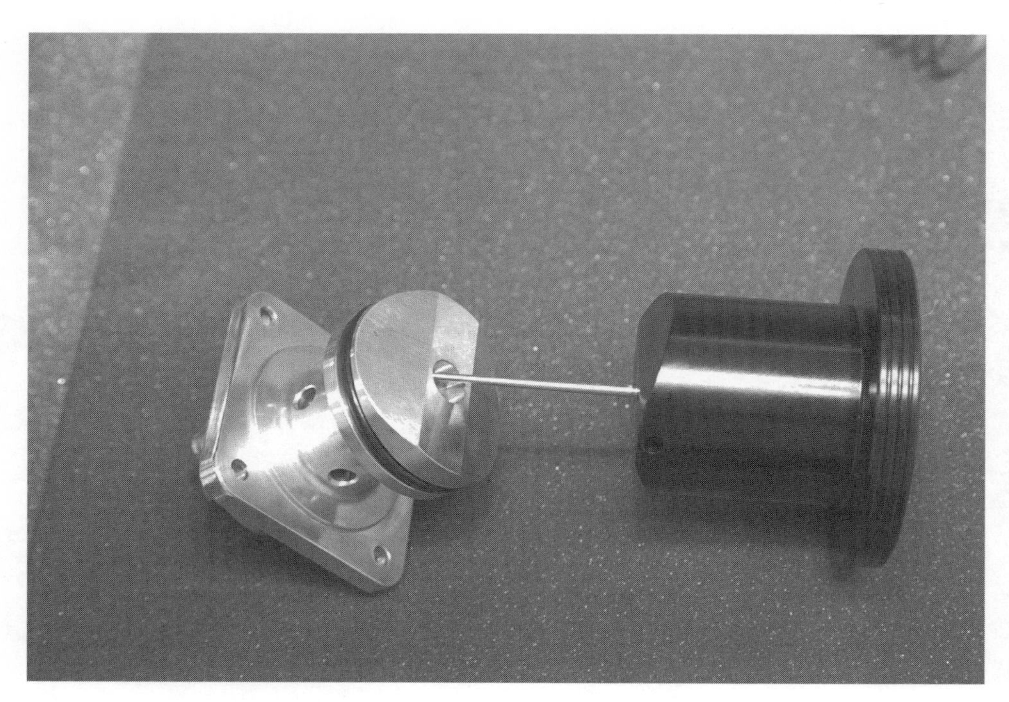

Here you see the slide with the tapered needle, and the adjustable main jet it fits into, from one of Carl's new Typhoon carburetors.

extra circuits needed with a fixed-venturi design and replaces them with one main jet and one tapered needle. Adherents of the CV carburetor design point out the fact that these carbs only open up to admit as much air as the engine can use under a particular load. You may open the throttle, but the piston will only open as far as needed. This keeps air speed through the carburetor high and aids throttle response.

Fans of the slide or smooth-bore designs like the fact that by eliminating the butterfly you eliminate a major obstruction in the carburetor throat and create the "smooth bore," a bore that will pass more air (with less turbulence) for a given size than any other design.

The Right One For You

The number of aftermarket carburetors available for the V-Twin engine is large and growing. Though a few are technically out of production, the SU for example, rebuilt models of those earlier designs are still available from reliable sources.

Whether your V-Twin is a mild 80 cubic inch street motor or a 114 inch monster motor there's a carburetor out there that's right for you, actually three or four carburetors. Once again the trick here is not to pick a "good" carburetor, but rather the one that's right for your application.

In choosing a carburetor for your aftermarket engine you need to consider how it fits the bike, will it interfere with your right leg and what style air cleaners will fit the carb. Some of the bigger carbs won't clear the heads and/or the five gallon Fat Bob tanks. Be sure when you buy the carb

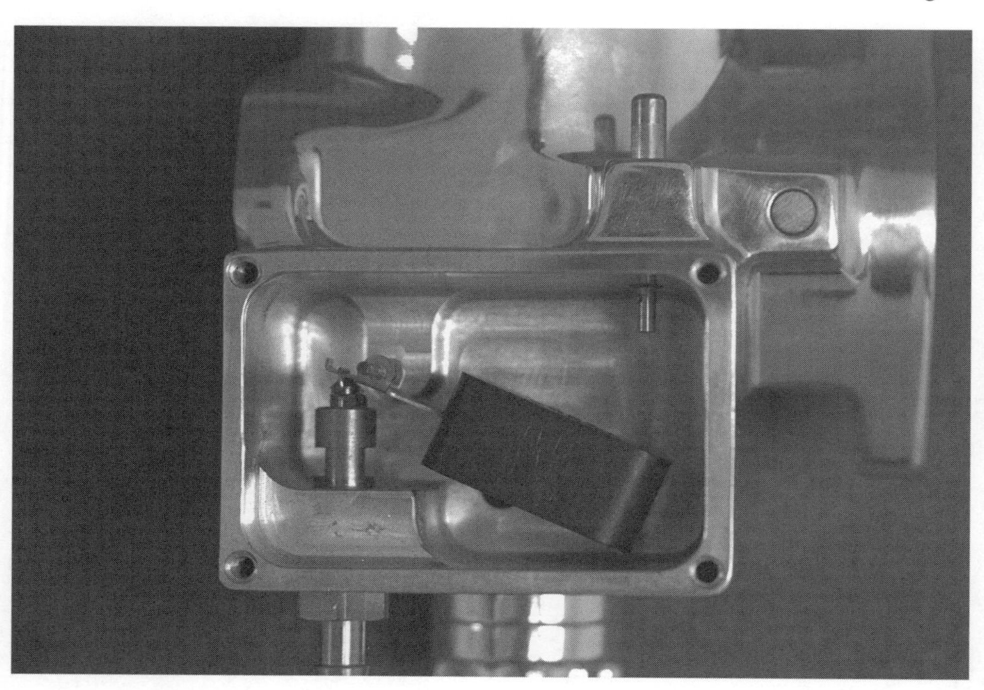

The needle and seat on a high performance carburetor need to flow enough fuel for high speed operation. It helps if the float bowl is easily accessible for checks and cleaning.

that you get the right style and length of throttle cables. If you want your carburetor in show chrome, note that some of the current aftermarket carburetors are available in polished or chrome plated versions, while other manufacturers insist you *not* chrome plate their carburetors. Check with the manufacturer regarding the availability of extra shiny renditions of your favorite carburetor.

The single most important thing to consider before buying the carburetor is how it will work in relation to the engine you've built. If you have a good shop doing all or most of the work on your engine consider their suggestions before buying a new carburetor. Ask around to be sure the carb you want is easy to tune and that jets and other parts are readily available. Remember bigger isn't always better and sometimes a carburetor that's too big results in low air velocity in the carburetor throat and poor low-speed response. This last comment is particularly true when considering some of the smooth-bore designs.

Because the emphasis of this book is street engines some competition-type carburetors have been left out of the following product descriptions.

What's Out There

The Typhoon from Carl's

Carl's new "billet" carburetor is a constant-velocity design, the only carburetor manufactured from a solid chunk of 6061-T6 aluminum. Inside the shiny aluminum body you will find a large piston or "slide" with a tapered needle pressed in at the bottom of the piston. Like all carburetors of this type, engine vacuum, working against the spring, determines the position of the slide. The position of the slide in turn controls the amount of fuel that enters the airstream as the tapered needle is pulled farther and farther out of the main jet.

This new carburetor measures 50.8 mm in diameter (two inches), big enough to feed the largest of motors. Yet this same carb works fine on a smaller 80

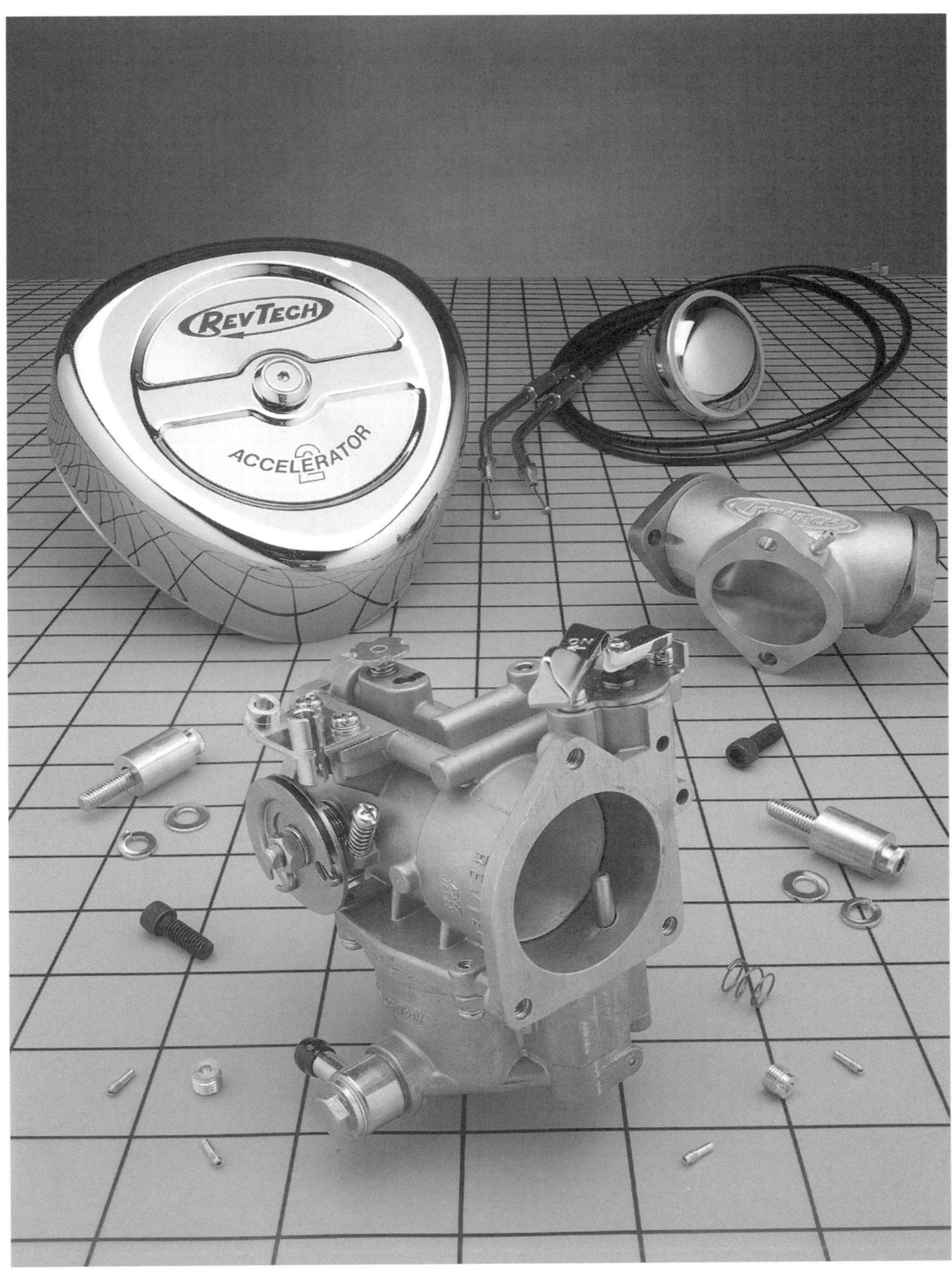

The RevTech Accelerator 2 carburetor is a butterfly design, with provisions which allow it to "grow" with your engine. Each RevTech carb comes with 3 different venturis (38, 42 and 45mm) so you can increase the size of the carburetor as you further modify your engine. Custom Chrome.

Professional builders encourage you to pick the exhaust system with an eye toward performance and not just aesthetics. 2 into 1 systems like this one from RB Racing tend to make good mid-range power. RB Racing.

cubic inch engine simply by changing the size of the tapered needle. The main jet can be adjusted externally without disassembling the carburetor, even while the bike is running, though if you ever do have to disassemble one of these you will find it to be very simple inside.

RevTech Accelerator II from Custom Chrome

Designed from the start for Custom Chrome, this butterfly-style carburetor features replaceable venturi-sleeves so the carburetor can grow with your engine. 38, 42 and 45mm venturi sleeves are available to help this carburetor adapt to everything from a 80 to 120 cubic inch Big Twin.

Most butterfly carburetors have an accelerator pump and this RevTech model features a high-volume pump designed to squirt the fuel into the center of the venturi for even distribution to the cylinders. Designed to operate with the current two-cable system, this carburetor has an enrichment device instead of a true choke. It allows the user to change both low and high-speed jets, and the high speed air jet for complete tunability.

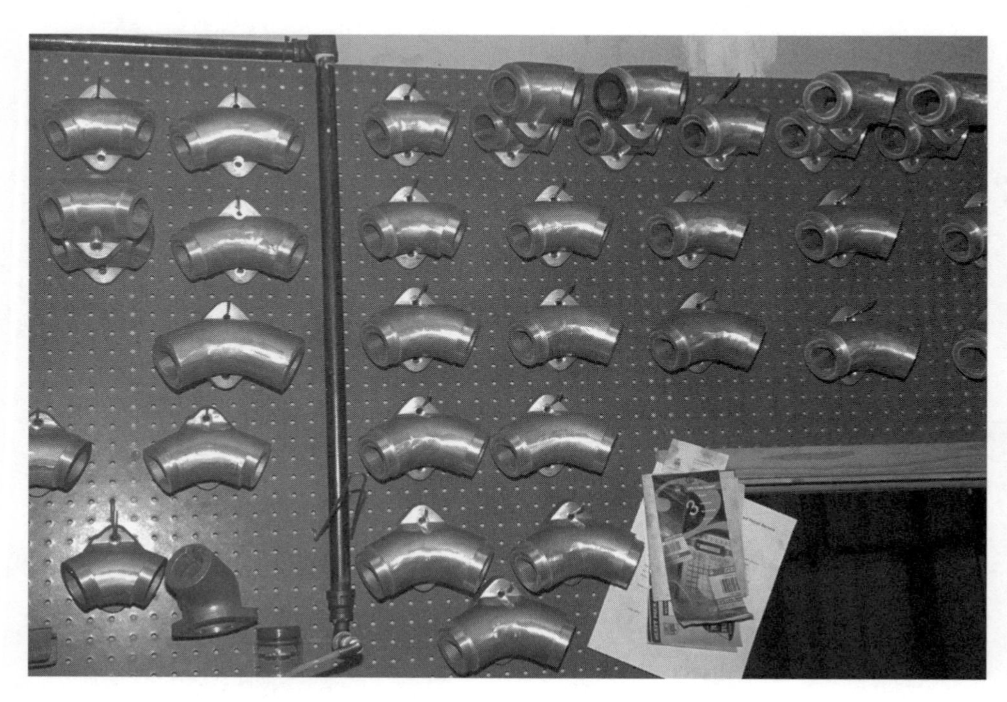

As engine size grows, so does the need for longer and larger intake manifolds. Here are just some of the manifolds Randy keeps in stock at Hyperformance.

QwikSilver 2 from Edelbrock

From one of the largest automotive aftermarket parts suppliers comes the QwikSilver 2 carburetor. The QuikSilver 2 is a flat slide, variable-venturi carburetor available in sizes ranging from 36 to 42mm. All sizes are available in two mounting styles, flange mount or grommet mount.

Some interesting innovations exclusive to the QwikSilver include a single fuel circuit controlled by a single needle that is the same width from top to bottom with a tapered flat section cut out of the back side of the needle. As the air passes through the venturi past the needle it creates a low pressure area which allows the fuel to move up the pickup tube and into the venturi where it atomizes instantly as it hits the airstream. As the throttle is opened farther by the rider the needle moves out of the pick up tube allowing more fuel to flow past the needle.

Another feature of the QwikSilver is the ability to self compensate for altitude changes. The float bowl is vented to the venturi rather than the atmosphere like most carburetors. By venting to the venturi the float bowl is pressurized with the same pressure as air traveling through the venturi.

Because the QwikSilver does not have an accelerator pump, venturi size is critical when choosing the right size carburetor for your specific motor. The recommended venturi size is based on cubic inches according to the following chart.

Recommended Venturi Size

55 to 73 cubic inches
36mm
74 to 84 cubic inches
38mm
85 to 103 cubic inches
40mm
104 and more...
42mm

Any questions regarding the QwikSilver should be directed to the QwikSilver division of Edelbrock (see Sources).

The SU carburetor from Rivera

Originally designed for use on automobiles like the venerable MG and Triumph, the SU is a constant velocity design that has long been used on V-Twin motorcycles. Like most CV designs the SU has both a butterfly valve and a CV slide in the carburetor throat. At idle the butterfly, connected to the throttle, is closed and the slide is positioned near the bottom of its travel. As the throttle is opened engine vacuum works to pull the slide piston up, which in turn pulls the tapered needle farther out of the main jet providing more fuel in direct proportion to the increased air flow.

The SU uses at least two interesting design details: A bi-metallic support for the main jet which provides automatic compensation for temperature change and

The SU carburetors have been around for a long time and are still in use today. A basic CV design, the SU provides the engine with only as much air as it can use for a certain speed and load.

hydraulic damping of the slide piston, which keeps it from moving too far too fast.

Available from Rivera and Rivera dealers the SU comes in kit form with the correct intake manifold and all necessary hardware, and is designed to work with dual-cable throttles.

Super E and G from S&S

The Super E and G carburetors from S&S are among the most popular of the many carburetors available for current evo-style Big Twins. These two "shorty" carburetors are both butterfly designs measuring 1-7/8 inch (47.6mm) for the E, and 2-1/16 inch (52.3mm) for the G ,measured at the throat. Compared to the earlier B model from S&S these two carburetors are 1-7/16 inch shorter and designed to tuck neatly into the right side of the bike.

When he's not building exotic street motors for clients like Jon Kosmoski from the House of Kolor or Jim Priesler from Chrome Werks, Jerry Wilhelmy assembles XR 750 engines for racers from around the country.

You can't take a picture of an intake port so when he has a cylinder head with ports that work good Jerry makes a mold with silicone rubber. It allows him to measure cross-sectional areas in the port and to "catalog" his designs.

Features of the Super E and G include an easy-to-reach idle mixture screw and changeable mid and high-speed jets. Instead of a choke an enrichment circuit is used. Attached to the air cleaner backing plate the "'choke" lever is located in a convenient position at the back edge of the air cleaner.

With two sizes available the S&S Shorty carbs will work on everything from relatively small displacement V-Twins to strokers and big-bore engines. The S&S Shorty carburetors are designed for dual cable operation. S&S also makes a 2-1/4 inch Super D for monster motors and race bikes.

Dell'Orto dual throat

If the S&S carbs are designed to tuck in nice and neat on the right side, the Dell'Orto dual throat is intended to do just the opposite. A two-throat butterfly design, the typical Dell'Orto dual-throat installation hangs way out there in the wind lookin' for air. New Dell'Ortos are no longer available though parts and manifolds are still listed in the Rivera catalog.

Despite the fact that these carburetors are no longer manufactured they still show up on custom bikes with regularity. Before buying one be sure you can find someone to help with tuning, which riders report is more difficult than with some of the more common designs.

Screamin' Eagle carburetor from Harley-Davidson

The Screamin' Eagle is a 40mm butterfly style carburetor designed for high airflow and increased horsepower throughout the rpm range. This carburetor from The Motor Company features an accelerator pump and three fully adjustable circuits. The Screamin' Eagle carburetor is designed to operate with a two-cable throttle system and comes with a high-flow air cleaner.

Mikuni

Long known as the manufacturer of high-performance fuel-mixers Mikuni has recently added more models to its line of slide-type carburetors. The new 42mm and 45mm carburetors are intended to answer the carburetion needs of large displacement V-Twins. The 45mm in particular is intended for all-out performance applications only. All three Mikuni carbs, they offer a 40, 42 and 45mm, feature a roller bearing slide assembly for precise movement and a light return spring. The new larger carburetors come with a larger capacity fuel bowl and an adjustable accelerator pump.

Like other carburetors of this type the Mikuni "smooth bores" have no butterfly in the throttle bore to restrict air flow at full throttle. This means that a 40mm Mikuni, the recommendation for mild 80 cubic inch motors, will often flow more air than a larger carb of a different style.

The Mikuni carbs are designed for a dual-cable throttle cable, most Mikuni kits come complete with their own cable set which may be used with a standard throttle assembly.

A final note: All the manufacturers I talked to in assembling this section stressed the importance of properly matching the carburetor to the particular engine it will be used on. Be sure to read all the manufacturers recommendations and call them before buying if you have lingering doubts. Try to buy your carburetor from a reputable local shop, one that can help with tuning tips or parts you might need for the installation.

Buying from a good shop will minimize the tuning you need to do after the installation. If tuning is necessary be sure to think first and swap jets second. If in doubt, don't be afraid to ask a few questions either from the shop where you bought the carb or from the technicians on the other end of the tech-line provided by most of the manufacturers.

A Tune-Up For Your Fuel Mixer - the Dyno Jet kit, Yost Power Tube and Thunder Jet.

Yost Power Tube

Available for the S&S super E, G and B as well as the factory Keihin CV carburetor and the 42mm Mikuni, the Yost Power tube aids fuel atomization though the use of their improved emulsion tube. The Yost kit also provides for more adjustment through the use of the additional jets supplied in some of the kits.

Dynojet

The Dynojet kit is designed to eliminate flat spots, improve throttle response, and aid in tuning the factory Keihin CV carburetor. Each kit comes with all nec-

An examination of these connecting rods led to a discussion of the metallurgy used in the rods, the pros and cons of using a bearing insert (as opposed to running the bearings right on the rod surface) and why parts are stress relieved before final machining.

essary parts including a new spring for the CV piston.

Thunder Jet

The Thunder Jet is a complete additional fuel circuit intended to provide extra fuel for mid range and high speed operation. These kits are available for most S&S carburetors as well as many other butterfly and CV carburetors. At least one company, Zipper's, sells S&S carbs with Thunder Jets already installed.

Fuel Injection For V-Twins

Despite the fact that nearly every new car on the road, and some new Harley-

The S&S Super E features a 1-7/8 inch throat and is recommended for smaller, milder V-Twins. The E's bigger brother, the Super G, has a 2-1/16 inch throat and is meant for larger displacement, highly modified V-Twins.

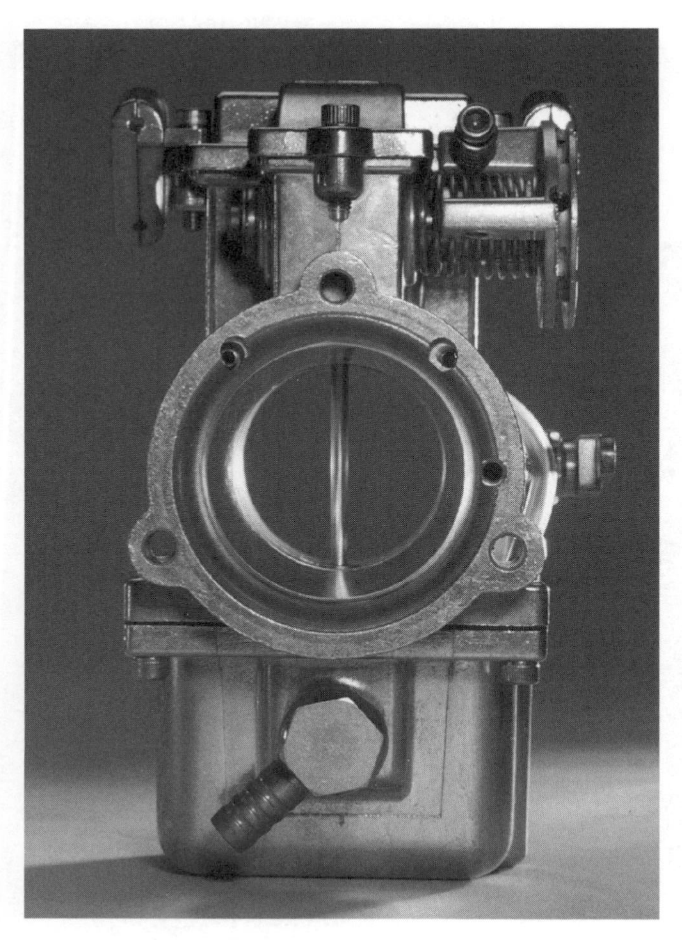

This shot shows the urestricted throat of a new QwikSilver carburetor from Edelbrock.

Davidsons, carries fuel injection as standard equipment, the world of aftermarket V-Twin motorcycles is still dominated by that old-fashioned device, the carburetor.

The trouble with fuel injection on V-twins is designing a system that can adjust, or be adjusted to accommodate the wide variety of engines equipped with an incredibly diverse bunch of camshafts and exhaust systems.

In spite of the obstacles fuel injection is sure to crossover to the V-Twin aftermarket. Thus you find here a primer on fuel injection and a description of three systems currently on the market.

A Little Theory

As mentioned earlier in the carburetor section the goal of any fuel delivery system is to deliver the right proportion of fuel and air to the cylinder in a combustible condition. A ratio of 14.7 parts air to 1 part gas is the ideal, or "stoichiometric" ratio. The engineer is faced with at least three problems when designing a fuel delivery system. First, the ideal ratio isn't always the same. Cold engines need a richer mixture, accelerating engines need extra fuel for maximum power. Second, for complete combustion the gasoline must atomize and mix thoroughly with the air. Third, the fuel must be carried by the air, meaning that the carburetor must be sized small enough to maintain good air velocity through the carb and intake. This is why you don't want huge carburetors on mild engines.

When it comes to delivering fuel to the engine in combustible form and meeting the three challenges outlined above, electronic fuel injection has a number of advantages over the carburetor. At the heart of most modern fuel injection systems are the injectors themselves. Gasoline must be mixed with air before it will burn. The mist created by the injection nozzles breaks the gasoline into very, very small particles. Smaller particles mix more readily with the air and are more likely to obtain enough oxygen for complete combustion, meaning more power and fewer waste products at the tailpipe.

If the fuel injection system can deliver this fuel mist just upstream from the intake valve we no longer need the airstream to carry the heavier fuel droplets. Throttle bodies can now be sized for maximum air flow instead of being sized to maintain good air velocity, and good throttle response. And with the injectors near the intake valve, when you whack the throttle open there is no waiting for the extra fuel to get to the cylinders.

Using a solenoid for an injector and controlling the solenoid with an electronic control module means great precision in delivering just exactly the correct amount of fuel. Input from various sensors allows the control unit to determine exactly how much fuel is needed under a certain set of conditions.

Most automotive systems use an oxygen sensor to help the system monitor its own fuel mixture, and operate in what's called "closed loop." Some of the aftermarket systems currently available for V-Twins have provisions for an oxygen sensor while others do not and operate in what's known as "open loop."

WhiTek Fuel Injection System

The WhiTek fuel injection system is the brain child of Bob White, from Arroyo Grande, California. If too many modern products are "designed by committee," Bob's fuel injection system is definitely the work of one man. Bob is currently the designer, production manager, and head of marketing.

The result of Bob's efforts to design a fuel injection system for a V-Twin motorcycle looks a little like a carburetor for the space shuttle. The system bolts on where the carburetor used to sit, but underneath the cover it doesn't look like any carburetor you've ever seen. Rather than have the control unit separate from the throttle body, the WhiTek system puts nearly all the components under the air cleaner. Packed under the cover are the throttle body with two fuel injectors, one circuit board, one electronic fuel pump and a wiring harness. There's one more component and it's one of the niftier features of this unit. That component is the monitor that bolts to the handlebars and plugs into the wiring harness.

The WhiTek unit is classified as a throttle body, Alpha - N unit. Meaning that the injectors are housed in the throttle body and that the major inputs to the control unit are from the throttle position and rpm sensors.

The WhiTek unit controls the ignition timing and advance curve as well as the fuel curve. Through the use of the monitor that mounts on the handlebars both the fuel and ignition curves can be altered by the rider. If you get a little pinging for example when you open the throttle on the highway you have only to set the module in the right mode, duplicate the "pinging" situ-

At Walters Technology they manufacture a "Pro-Back" for the Mikuni 45 (and the 42) which aids atomization and throttle response when the carb is near wide open throttle.

From the front the WhiTek fuel injection system looks like a standard carburetor.

ation, and then dial in a little retard when the pinging starts.

The extreme adjustability of the WhiTek unit is touted as one of its strengths. From a stock 883 to a full-on stroker, this unit will deliver fuel and ignition, single or dual-fire, in the correct amount and at the right time. Simplicity is another strong suit. Though some would suggest that speed-density units using rpm and vacuum sensors allow the computer to better sense the engine's load, Bob insists that his system is simpler and also allows you to run nearly any camshaft, even one that results in very little vacuum at idle.

E.T. Performance

With a background as an aerospace engineer, long time motorcycle enthusiast Ted Shrode is in a good position to design and build his own fuel injection system. Ted's system uses the ECU to control both the fuel and ignition maps and allows the user to graphically alter either curve to suit his or her unique needs. This feature allows the E.T. system to work on stock 80 cubic inch V-twin just as well as it does on a turbocharged drag bike.

The ECU used for the E.T. system stores data from a run or road test for analysis later. With an oxygen sensor the system will run in closed loop, without the sensor it runs open loop. The complete kit includes intake manifold, throttle body, fuel pump and injectors, CPU, and all hoses and wiring.

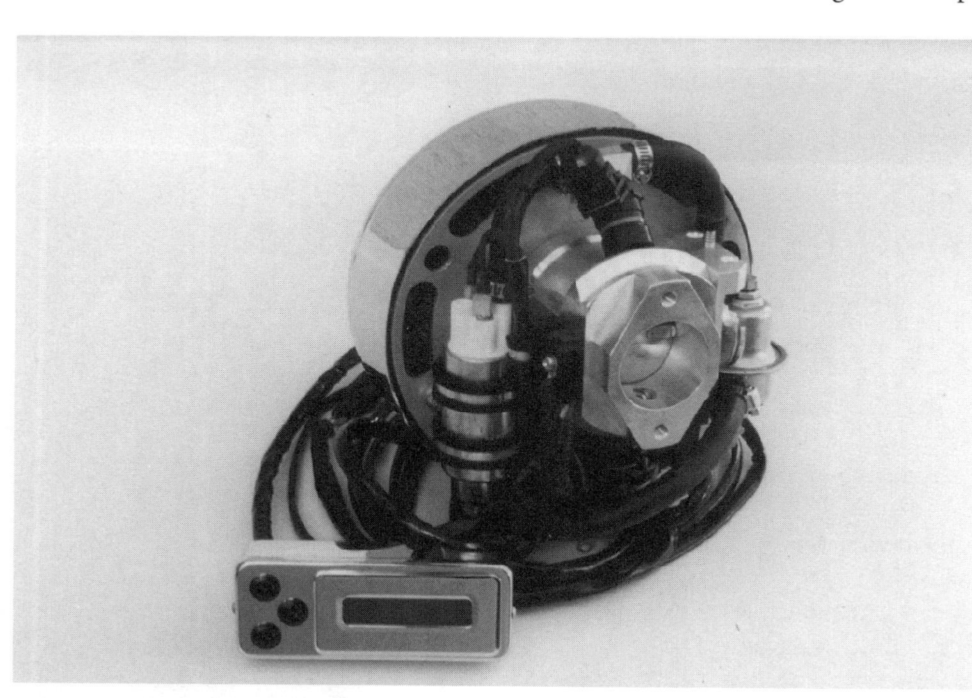

Yet, behind the cover lurks the throttle body, fuel pump, injectors and processing unit. New sensors allow for the display of oil temperature, oil pressure and inlet air temperature. The WhiTek even has a automatic spark retard feature for bikes running NOS.

RB Racing

RB Racing, the company that builds 400 horsepower Ninjas (for the street!) and 200 mph BMWs for Bonneville, has a fuel injection system tailor made for your V-Twin.

Marketed under the RSR (Race Systems Research) label, this system includes one of two possible throttle bodies which house 2 or 4 Bosch injectors. The injectors are controlled by their own ECU which receives inputs from manifold absolute pressure, throttle position, air and oil temperature, and oxygen sensors.

The ECU comes with their own Autocal software which gives digital or analog readouts of air fuel ratios and engine load factors. With this software the user can edit the entire program point by point or let the ECU write the maps, which you can edit later through the use of their optional, on-board readout device for air-fuel ratios. Programming and editing of the fuel maps can also be done with a PC hardwired into the ECU.

The complete kit comes with the ECU and harness, all sensors, which can be found at most auto stores or GM dealers for replacement, idle air control motor, throttle body, manifold, air filter, fuel pump and regulator, lines injectors and hardware. Optional is the Autocal software which simplifies the programming of the fuel curves.

Shop Tour: General Engineering Inc.

Jerry Wilhelmy is a man highly regarded for his ability to assemble very special engines. From racing XR 750s to supercharged V-Twins for the street, Jerry builds motors his way. As Jon Kosmoski, owner of House of Kolor, puts it, "Jerry is beyond the simple mechanics of things and into the engineering side, the essence of making things really work."

Jerry is a man who likes what he does, who puts in long days and often nights doing work of a very high caliber. If you ask Jerry to lunch he explains that he doesn't have time. I did, however, manage to ask Jerry just a few questions about the essentials of horsepower and torque as they apply to a V-Twin engine.

You made the comment that: "Engine performance is all in the heads, everything else just has to go up and down." Can you; 1) expand on your comment, and 2) comment further on the importance of the carburetor and exhaust in light of the comment?

The efficiency of any internal combustion engine

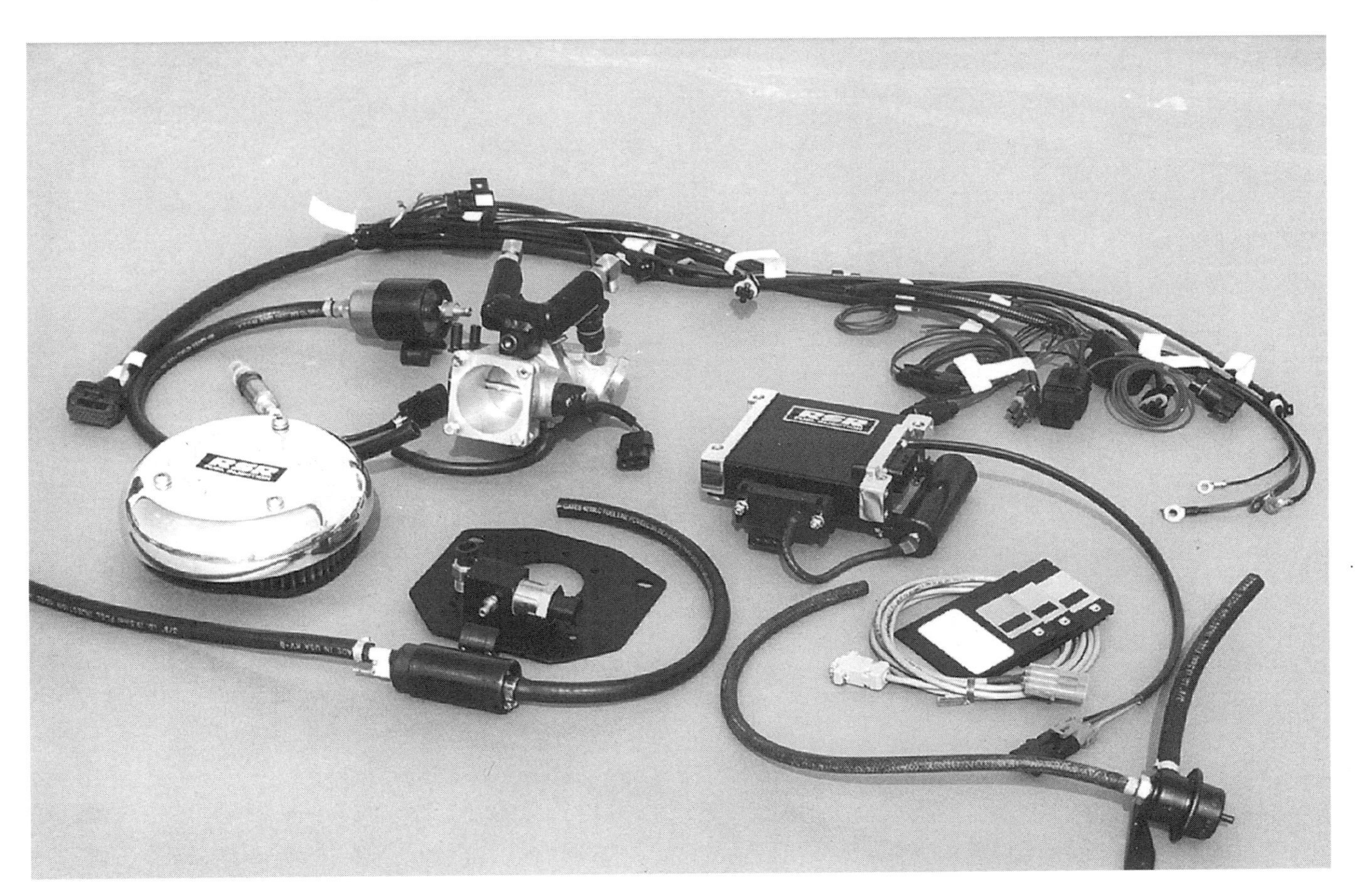

From RB Racing comes this system (with variations) to fit nearly any Big Twin. Along with their own CPU the RB folks have developed an automated calibration software package to aid the user in tuning the fuel injection to run on nearly any V-Twin.

is directly related to its ability to breathe. The mechanical limits of the engine components must be matched to the rpm which the cylinder heads permit.

When people go looking for heads (or have their own heads ported) they're looking for big flow numbers, yet there is a need to maintain good velocity through the ports. Can the ports be too big, on a street engine? If so, how does a head designer, or engine builder modifying heads, balance the need for flow with the subsequent need for good velocity?

Good ported heads will be matched to the engine rpm and piston area. The mean port velocity at the toque peak should be 240 to 250 FPS (feet per second) The mean gas speed at the HP peak should be 350 to 360 FPS. Other factors can affect these number such as rod-to-stroke length ratio because this affects the gas speed greatly at point of crank rotation which are critical due to the changing piston velocity. Port areas and the velocity number listed above would be different in a short rod motor. Each different motor combination would require slight alteration in this area.

With regard to the engine's internals you made the comment that, 'What's round needs to be round and what's square needs to be square." Can you expand on that comment to help the rest of us understand what you think is important about the machining and set up of a good V-Twin.

Harley-Davidson engines to most, appear to be simple and quite easy to understand. While the design is really complex and difficult to manufacture or repair. A good engine builder must have the full knowledge of a machinist and inspections equipment. The design is further complicated because of the air cooling.

You said "cast pistons are better than forged," Or maybe you said you prefer cast pistons in street engines. If the forged pistons are stronger, why do you prefer the cast pistons, is it a matter of fit and clearance?

Cast pistons have the advantage of being cheaper, lighter and expand less, which permits closer running clearance. This benefits ring life and noise reduction. Unless a motor is prone to detonation or pre-ignition cast pistons will almost always be adequate.

There are advantages to running one carburetor per cylinder, but the only V-twins normally run that way are the XR series Harley-Davidsons like this XR 1000 of Larry Page.

When I asked about "combinations" of parts that make good power, you said you like long exhaust pipes, and stock-style header pipes with the cross-over pipe in place. Do you have similar personal guidelines for camshafts and carburetors that work well on high performance street bikes.

Most street Harley motors require long exhaust pipes to aid in the tune because of the low rpm which they are tuned for. The factory cross-over pipe has two good advantages: One being that the small mufflers used on the motorcycles can be restrictive, the balance pipe allows equal pressure and exhaust flow through both mufflers. The cross-over pipe also provides some control of the returning wave and prevents some EGR (exhaust gas recirculation).

My personal guideline for selecting components is basic. You must know the true needs of the customer. Most talk HP but really want torque. Most want more HP but not higher rpm. HP = Torque X rpm/5252.

If you don't rev it the only logical way to make more HP would be to enlarge the motor and keep it on the same state of tune. The stock bore and stroke 80 cubic inch V-Twin can be tuned to have 50 HP or over 100 HP and both run on pump gas. But the 100 HP motor will need to be tuned to over 6500 and the torque peak would be higher than the HP peak of the 50 HP motor. V-Twin motors as designed are best suited to be tuned for torque because of the long stroke, small bore and inefficient head and valve train. END

INTERVIEW, CARL MORROW FROM CARL'S SPEED SHOP

Carl Morrow of Carl's Speed Shop is a man well versed in the pursuit of speed. Holder of many records, this is the man Buzz Buzzelli, well-known writer on V-Twin topics, called, "A long-time drag and salt flats competitor with an endless string of successes"

Carl, let's start by talking about the parts you like to use on a typical high-performance V-Twin for the street. What combinations you use. How you pick a carb. How you pick a cam. What about heads. All of those kinds of things.

It's real simple. At my experience level, what's successful we continue to use. And we only tune two carburetors. Our own Typhoon and the S&S. And we have a great amount of success with both.

What makes a good carburetor?

A good carburetor is flexible and it's tunable. It will run through the entire rpm range with flexibility and it builds good horsepower.

Let's talk a little about your new carburetor and why you designed the carburetor you did. Why is it a CV (constant velocity) and not either a straight slide or a butterfly style carburetor? What is it you like about the CV?

It actually is a butterfly and a CV. My carburetor, how I designed it, it is operated by the vacuum of the engine. If you're at a low rpm level and you're underneath the cam, so to speak, or your exhaust system, you could open the throttle wide open and the slide will not raise. It will be a small carburetor until it becomes a big one as the engine needs it. The slide will keep it smaller until the engine rpm gets up to such a point that it will take the full size of the 50.8mm or 2 inch opening. So having that, it's more flexible in higher altitudes, because if it doesn't have the air, it won't lift the slide all the way and will not give it a richer mixture.

So it's self-regulating?

It's self-regulated. Being self-regulated like that it has instant response like no other carburetor, for that

The Keihin carburetor used by the factory is often bypassed in favor of sexier and "faster" designs. Yet, because the Keihin is a CV design measuring 40mm, it makes a good carb for engines closer to 80 cubic inches with fewer modifications - and you can find them cheap at the swap meet.

very reason.

So if you whack it open it gives the motor exactly what the motor needs.

Yes, it does. Mine is so simple, it only has two adjustments. On the bottom, it's a 7/16 hex nut and you turn it one flat a time, richer to the left, or to the right to lean the mixture. This carburetor only has one jet orifice.

There's a tapered needle that fits into the jet?

Yes. And the jet orifice is moved away from the tapered jet needle to richen the overall mixture through the full entire range. You get back to the same thing, if the motor doesn't take the mixture, the slide will not run. Unlike other "slide" carburetors that have a cable hooked to the slide, where it opens up manually and the carburetor has to run to catch up.

Would it be fair to say that the new Typhoon is a similar design to the old SU?

The engine in this Dave Perowitz bike features an SU carburetor and a magneto ignition.

No, I completely changed it and I have a patent on it. It is only similar in that it is a vacuum-operated slide, and the similarities would end there.

If I'm a guy building a motor and I'm going to go buy a carburetor, what are the kinds of parameters I need to consider before I put down my money? There are at least five or six brand names that are well advertised. What are the considerations I need to use as I try to pick the right one?

How successful is the manufacturer? Is the company successful in advertising or performance? If you're putting a performance carburetor on, I'd look to the track record of the individuals that have manufactured the carburetors to see what kind of results they can really hang their hat on.

So it makes sense to buy the carburetor from a shop that can give me some help tuning, if in fact, I need that?

Certainly. You want to have a capable shop in your area or have support from the manufacturer to make those problems, if they do arise, easy to deal with. Absolutely. And the simpler the carburetor the better.

Do people get in trouble when they tune that carburetor? Whether it's yours or an S&S butterfly carburetor? Do people chose poor procedures and poor diagnostics when they are trying to straighten out a carburetor that isn't tuned right?

I think a lot of people need professional help when it comes to tuning a carburetor. When you get a carburetor from someone who knows what to do, however, it should already be jetted correctly for your application no matter what brand it is.

So it comes back to buying it from either a good retailer or a good shop with a lot of hands-on experience.

I think, to begin with, that anyone dealing with performance should really look for references of how that shop operates. If they're just a shop that has someone working at the parts counter who is going to sell you anything that you'll buy because it's profitable for them, then you want to go somewhere else. You need someone who has a reputation for understanding how to tune Harley-Davidson engines. And that reputation alone will follow through on the carburetors they choose to use. If they have integrity they won't let the customer make a bad choice just because they're going to make a commission on it.

I'm talking Evo Big Twin here, but there is still more than one intake manifold. Does a person have to worry about the style of intake when buying a new carburetor?

Yes, that's a consideration. It's better using the O-ring clamp style. It seems to be the less troublesome.

Another thing too, our manifold has been carefully designed and flow-tested, it doesn't need any work done to it. We feel that boring our manifold would not help it at all. We spent a lot of time developing that manifold and in fact when you buy one of our carburetors you also get one of our manifolds.

In the case of a butterfly carburetor, you can end up with a carburetor too big for your motor?

Yes. Easily.

When people come into your service department, do you see some of that?

We see a little bit of everything. People often come in with this carburetor, they jetted it too richly, and they've got a set of straight pipes on the bike. That is just one of the hardest things to correct. Some have two inch straight pipes with baffles in them, and that is just double bad. The baffles don't help the straight pipes, it's just like stuffing part of a potato in there. It's just ridiculous.

If I'm looking for a shop to do some or all the work on my engine, how do I find a reputable shop?

You need to check references. Track records. When you're doing this kind of work, no matter who you are, you are still only as good as your last job.

I see more and more people selling kits. Heads, intake, carburetor, cam, exhaust sometimes. It's more dollars for the consumer, but if the components are chosen correctly, it seems like a good way to go?

Well, that's what the buyer doesn't know. There again, you've got to look at what are they hanging their credentials on. What is their background? (the shop selling the parts). Are they just merchandising people that rely on a lot of colorful ads and their own dyno-comparisons? Or do they back up their claims with some track experience

We have kits, our number one, two, three and four "Get Kits" for 80 inch big twins. We're careful that we don't oversell the customer, we give him just what he needs, what the rider can use. Most customers are like hungry people going to a buffet. They take all this food on their plate, and they can't eat it. They do the

Though many carburetors come in a kit with the recommended intake manifold, aftermarket replacement manifolds are available with the compliance fittings. Most (not all) engine builders prefer the compliance design because it allows the cylinders to grow up and apart as engine temperature increases. Custom Chrome

same here, if you let them, especially the ones that have the big budgets. We like to give the customer just what they need at a very good budget price, and they will grow into it. If the customer needs more, he will buy more.

Good concept. Do people sometimes want to apply full race techniques to street motors? Do they get in trouble that way?

What they really do when they come to me, they've read in the magazines all these ideas of what ignition they need and what new whiz bang hot parts are the best to have, and they put a motor together using all the sexy new parts and it turns out to be a cluster mess. It couldn't be tuned by me or anyone else. But they give me a phone call and explain that they bought all the best parts and now they just want me to "tweak" it. That's the buzz word they're using now. When I ask them, "what does tweak it mean?" Well, they'd like me to set the timing and adjust the carburetor and jet it correctly. Then they believe that by doing these small items it will make that whole package come suddenly alive. And I have to back off because it's not true.

So then when I tell that person, you need this change and several more changes they look shocked and wonder why that is because they already have the best parts. That's a real problem. A common mistake of ordering all the wrong parts. People need to work with someone who will guarantee results - so that they can get the horsepower that was promised. That way they get what they paid for.

If they had spent a little bit more money on the front end with a qualified shop, then they'd have a lot more in the end?

No, it's not more money. Here, or at a good reputable shop, they'd spend less; because they'd spend it better. They would have the correct parts. For instance, they are often buying some ignitions and some different parts that don't need to be added. Inexperienced people think that the more complicated it is, the better it is. That isn't true. It's always been my motto, all through the years, 'never use two when one will do.'

If you mill the heads, add longer-than-stock cylinders or install heads with raised intake ports, you will likely need an intake that's wider or narrower than stock. Both are available from Custom Chrome in spigot styles to accommodate Keihin or Mikuni 42mm carburetors. Custom Chrome.

Do you do many nitrous installations?

Not if I can help it. I don't want it. Do not like it. It destroys engines. Nitrous destroys engines, period. It's a fuel. Everybody knows when you start accelerating horsepower with a fuel it's hard on the engine.

So you've got the same engine component but under double or triple the stress it was designed for?.

Something along that line. Probably triple. We've used it occasionally. We've used it with the Aerocharger. We got high horsepower with it, but I'm not a fan of it. Basically you use it in self defense because all those guys are pouring bigger loads of nitrous in their motors.

Are a high percentage of your motors 80 cubic inch motors? And at what point do you recommend people move up to a 96, like an S&S?

Yes, we do build a lot of 80 inch motors. In terms of the bigger engines, it depends on what the owner wants and things like how heavy the bike is. One of our favorite larger engines is an 88 incher that is dynamically balanced with head work cams, carburetor, ignition and exhaust. With that combination we get 97 horsepower at the rear wheel. Those are real happy motors that live. They're stronger than stock because of the S&S cases, number one. The combination works out real well. If somebody wants to pull more weight and has a desire for more torque, then for a road motor we'd use a 96 inch S&S motor or 97 inch Axtell-cylinder motor with a big bore and a stock stroke. They're very good.

But most people don't realize how much performance we can get out of an 80 incher. For the multitudes, the 80 inch motors are the way to go.

Carl Morrow from Carl's Speed Shop leans on a bike in his new facility located in Daytona Beach, Florida.

Chapter Six

The Ignition System

Light the fire

How It Works

Before jumping into a full blown discussion of the various ignition systems available for your bike it might be helpful if we first took a brief look at how a basic ignition system works.

First, a look at the coil, which is really just a transformer made up of two sets of wire windings, primary and secondary. If the coil is wired into a simple points-style ignition system current runs from the ignition switch to the positive terminal on the coil, through the

A wide variety of ignition systems are available for that new V-Twin. Some use a magnetic pickup and a separate module while others combine pickup and module into one little black box that mounts under the points cover.

primary windings, out the negative terminal and then on to the points. If the points are closed the current flows through them to ground. This current moving through the coil's primary windings creates a magnetic field which surrounds the coil. When the points open the current in the primary circuit is interrupted, which causes the magnetic field to collapse onto the secondary windings.

As you no doubt recall from Science class or Physics 101 when a magnetic field passes over a conductor a small current is induced in the wire. In the coil a strong magnetic field passes over thousands of "wires," resulting in a sharp voltage rise within the secondary windings. In essence, primary current of just a few amps and 14 volts is transformed into a 30,000 volt spark with amperage measured in miliamps.

The Dyna 2000 module interfaces with either the factory pickup or the Dyna S pickup assembly. A very flexible module, the 2000 will work in either single or dual fire mode, and has 4 different advance curves.

All points type systems have some basic problems. Namely that the rubbing block is always wearing, changing the timing and reducing ignition output and that the points themselves become pitted and dirty thereby reducing current flow in the primary side and output on the secondary side. The additional drawback is the mechanical advance unit used with points-type ignition systems, which are not as reliable and are harder to adjust than the modern ignition curves pre-programmed into the various ignition modules.

In nearly all modern vehicles the points have been replaced by some type of electronic pickup. Most modern vehicles, including factory Harley-Davidsons, have replaced the points with a hall

From Harley-Davidson comes the Screamin' Eagle ignition module which replaces the stock rev limit with a 8000 rpm limit.

effect or magnetic sensor which is often connected to the ignition module. By replacing the points with some kind of sensor you eliminate the multiple problems of worn rubbing blocks and pitted points.

Now instead of the points opening as the piston nears TDC, the pickup "senses" the rotor as it spins by and signals the module to open the primary circuit.

Factory V-Twins use the module to advance the spark, based on rpm and whether or not the VOES (vacuum operated electrical switch) is open or closed. This VOES is normally open under acceleration and conditions of low vacuum and closed during idle and cruise, or high vacuum, conditions. Each factory module has two advance curves, a faster curve for optimum conditions and a slower curve with less advance to minimize pinging under heavy load conditions. In this way low vacuum conditions signal the module to use the "retarded" curve.

Aftermarket Systems

All ignition systems use a primary and a secondary side. Differences in the aftermarket systems include the type of sensor used to provide timing information, whether or not the system uses a module, how they control the ignition advance, and whether or not a VOES switch is used. Some systems use the factory pickup and simply replace the stock module with one with a faster advance curve, or a curve that can be tailored to suit your needs and riding style. Some of these modules use the VOES switch while others do not.

Some aftermarket ignition systems use their own ignition pickup and eliminate the module all together. Some of these rely on mechanical advance weights while others use the module or 'black box" to control the ignition advance.

The HI-4 from Crane combines the pickup and module into one small unit which mounts under the points cover and handles all spark generating, timing and rev limiting chores.

Single and Dual Fire

Factory V-Twin ignition systems and many aftermarket ignitions operate in what is known as dual-fire mode. That is, the coil fires both spark plugs at the same time. Stated another way, when one cylinder is near TDC on the compression stroke and ready for the power stroke, the plug on that cylinder fires, at the same time the other cylinder also fires. The spark in the "other" cylinder is known as the waste spark. Specifically, when the front cylinder fires the rear cylinder is at 10 degrees ATDC; and when the rear cylinder fires the front cylinder is at 80 degrees BTDC. These figures assume 35 degrees timing advance.

In theory it would be more efficient to fire each plug alone, called single fire, and the more sophisticated ignitions do exactly that. When the front cylinder comes up to TDC that spark plug, and only that spark plug, fires. Though before-and-after dyno tests show little or no direct increase in horsepower (at least the tests I have seen) the switch to single-fire ignition always seems to result in a smoother running engine as judged by seat of the pants testing.

Ignition Buyers' Guide

The ignition system you install can be a "traditional" model with pickup, module and coil; or one of the newer systems that eliminates the module and mounts everything but the coil under the points cover.

Most builders of scratch-built bikes use a factory harness and most new aftermarket ignitions are designed to plug right into such a harness. Some builders stay with a factory pickup and either a factory or aftermarket module simply because the parts have been proven reliable. And if there is any trouble replacement parts can be found at any dealership and many independent shops.

Before you buy remember that electronic ignitions put out more power, more reliably, than any points type system. Despite the allure of "simpler" and older systems, the hot ticket is an electronic ignition with electronic control of the advance. Which one will depend on your budget and your engine. More compression requires more spark to jump the same gap. Getting enough spark means a reliable pickup, the right module (if your system uses one), a coil with enough output and wires with enough quality to deliver that output to the plugs without any danger of leakage or crossfire.

Note: No matter which ignition you buy be sure to follow the directions, use of the wrong ignition wires (many insist you use resistance wires and not solid copper core wires) can damage that expensive new ignition system.

Accel

The Accel Mega-Fire ignition module will plug into the factory wiring harness in place of the stock module and offers adjustable advance curves. Available as the single unit or in kit form, with Accell Super coil, 8.8mm ignition wires and U-groove spark plugs, the Accell module works with the factory pickup.

Compu-Fire

Compu-Fire makes a variety of ignition systems, some of which rely on a mechanical advance unit while others use an electronic, and in some cases adjustable, advance unit.

Most models operate in single or dual-fire mode and many have an adjustable rev limiter. The Elite series Compu-fire ignition systems come complete with coil and spark plug wires, and are available to fit everything from one-plug per cylinder dual-fire engines to two-plugs-per-cylinder single-fire applications.

Crane HI-4

This ignition system mounts directly to the ignition timing plate for ease of installation and trouble free operation. This self-contained unit requires no additional ignition module. Available in single-fire and dual-fire versions, in off-road and EPA legal models, this ignition system features fully adjustable advance curves, kick or electric start, a tachometer output, and a rev limiter that is adjustable from 4000 to 8000 rpm.

Dyna

Dyna offers a number of ignition systems for V-Twins including the Dyna 2000, a programmable ignition module which plugs into factory harnesses and replaces the stock module. The 2000 has the advantage of adjustable ignition curves and rev limiter and can be operated in single or dual-fire mode.

For competition use Dyna offers the 4000 as a complete kit with a 4000 series module, high output coils (single or dual-fire) and high-energy plug wires. Designed to be used with a Dyna ignition pickup the 4000 uses a two-stage rev limiter for easy staging at the drag strip.

The Dyna "S" is a self contained system which mounts under the ignition cover. Designed to be used with a mechanical advance unit, this Dyna system is available in either a single or dual-fire model.

Harley-Davidson, Screamin' Eagle

The Motor Company makes a variety of good modules that are often used to improve the performance of a high performance V-Twin. Most commonly used is the Screamin Eagle module which replaces the factory rev limit with a 8000 rpm top speed and uses a fast advance curve. The other module in the Screamin' Eagle catalog is the '86-'87 Sportster performance ignition module (32420-87B) which also uses the 8000 rpm limit but with a stepped advance curve. Note, you may want to use an additional rev limiter rather than rely on the 8000 rpm limit of these modules.

Also available from your local Harley-Davidson dealer is their performance coil with 40000 KV output and a faster rise time.

RevTech

The RevTech Digital Electronic Ignition "uses digital technology to enable you to adjust it for the specific configuration of your high performance engine."

This RevTech ignition module plugs into any late model factory wiring harness, and features four possible advance curves and four rev-limiter settings. Designed to be used with the factory ignition pickup this ignition system can be used for dual-fire or single-fire operation.

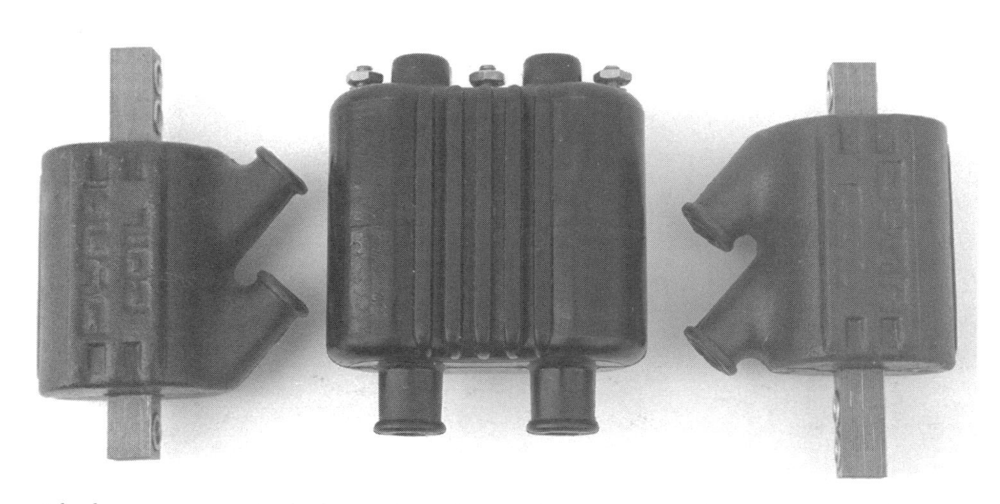

Whether you run a single-fire or dual fire ignition you need a high quality coil(s) with enough spark energy to fire a high compression engine.

Chapter Seven

Visual Considerations

Jewelry on the outside *and* the inside

The Money You Spend On The Outside

For most readers the Ultimate V-Twin Engine is part of an Ultimate V-Twin Motorcycle. The total expense is considerable and the engine is a big part of the total dollars spent. The engine is also a big part of the looks for that new motorcycle. So as you assemble a plan for the Ultimate thumper consider not just the jeweled goodies that go inside the cases and cylinders, but the cases and

This Donnie Smith bike is a good example of making the engine part of overall design for the bike. Note how the engine parts are either polished or painted to match the bike. Even the primary drive and clutch follow the same theme.

cylinders themselves. Should they be polished or plated, painted or powder coated? Don't forget to add the cost of these "extras" to the budget for the engine.

The engine is not only the biggest single expense for the new bike, it's also the biggest single visual component of the motorcycle. What makes American V-Twin motorcycles great is their mechanical-ness. The fact that everything is out there for all to see. The cylinders and heads aren't hidden behind a panel - they're right there in front of God and everybody. Whatever you do, or don't do, to the outside of the motor will have a major impact on how that motorcycle looks.

Not so long ago a wrinkle-back engine with lots of chrome-plated goodies would have been enough for almost any show bike. Today the best looking engines are painted and polished, adorned with tasteful amounts of chrome and polished billet. The best looking engines are designed as part of the overall motorcycle. Many are painted in a color that matches or compliments the paint on the bike. The polish and chrome are accents to the paint and overall design.

At the new Arlen Ness powder coating facility the "paint" is actually a finely ground powder, "applied" electrostatically with a very unusual spray gun. Carmina Besson

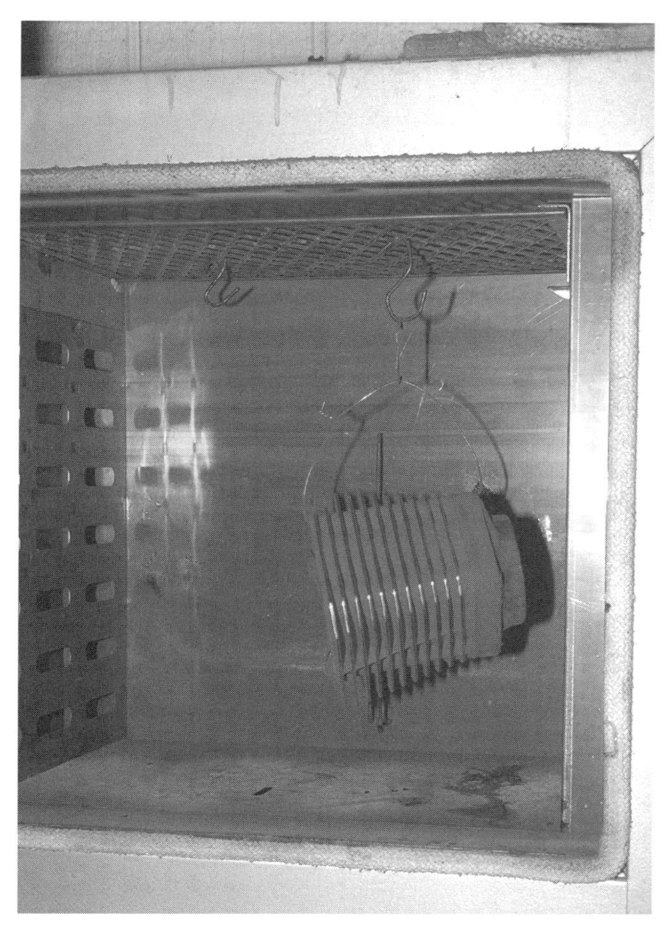

What actually causes the paint to bond so tenaciously to the metal is the heat from the oven. The powder paint can be carefully wiped off areas where it isn't wanted before the bake cycle begins. Carmina Besson

Most of the engines will be assembled from parts, which means there is no reason not to take the time to do some extra paint and polish work. Some shops buy complete engines unassembled simply so they can send cylinders out for painting and/or polishing without having to disassemble the engine first.

After the engine is fully assembled and installed it's too late to polish the fins. You need to paint and polish the engine before it's installed,

Shop Tour: Deters Finishing

Joe Deters begins the polishing process with the edges of the fins held up against the 120 grit belt. The next step (after doing the recessed areas) is to switch to a worn 240 grit belt. Each material needs to be treated differently, the sequence Joe uses here is best suited to cast aluminum parts.

A die grinder equipped with a cartridge roll is used to get down into the concave areas. The cartridge rolls come in the same grits as the belts. Joe starts with a 120 grit and moves to 240 (often used with lubricant to soften the effects of the abrasive) before going on to the finishing roll.

The cone buff is used to finish up the recessed areas. Joe uses a "cut-and-color" compound first and then finishes with a coloring compound - to bring out the luster.

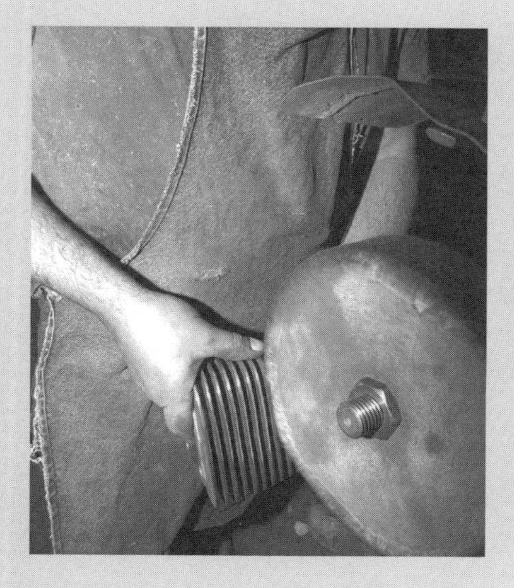

After working most of the finned area with the worn 240 grit belt, the edges of the fins are polished with a buffing belt (not shown). The final step is this soft buffing wheel, first with cut-and-color compound and then just coloring compound.

Though you can clean the parts with thinner, Joe likes wax and grease remover - it leaves no residue behind. He is quick to clean the cylinders while they are still warm and in fact likes to do all the polishing and buffing on warm aluminum.

The finished product, ready for masking and then paint. According to Joe, success lies in not using anything more aggressive than 120 grit, and making sure that each step fully removes sanding marks left by the step before.

and you need to decide exactly what you're going to do before that. Don't miss the opportunity.

Design an Engine

Maybe it goes without saying, but you need to design an engine that's an integral part of the motorcycle in a design sense. The trend is toward engines that combine paint with polish and/or chrome parts. Your job is to design an engine that looks *right* sitting in the chassis.

So you're probably going to paint the engine. Bare aluminum engines have become uncommon. Why not give the color and overall paint scheme some thought instead of just using the "default" setting.

When you pick out accessories consider that less is sometimes more. Chrome and polish used in conjunction with paint sometimes provides more contrast than a situation where everything is super shiny. If you don't have enough ideas of your own, ask the painter for his or her ideas. Or check out the magazines. From *American Iron* to *Easyriders*, each one is jam packed full of great custom bikes. What are those builders doing with their motors. How do they make the motor add to the visual impact of the motorcycle? Consider all your options.

What Can You Do

Paint

Engines can be painted in much the same way you would paint a fender or gas tank. To quote Jon Kosmoski from the House of Kolor, "The most important thing is to get everything clean, really, really clean." Once you've got everything clean, you need to use a good, two-part primer"

If the parts have an oily coating hot water and dish detergent make a good first cleaning step, followed by lacquer thinner followed by a Prep Sol type of cleaner that leaves no film behind.

This is the finished product, new cylinders hexed, polished and powder coated. Arlen Ness Inc.

The best primers to use on aluminum are those that contain Zink chromate. KP-2 and EP-2 from House of Kolor are two such paints, others are available from most major paint manufacturers (Zink chromate levels are dropping in some products due to environmental concerns, so you might want to check with your local paint jobber).

Because the engine is subject to heat and chemicals the topcoat paints should be catalyzed urethanes. These paints are much tougher and resistant to rock chips and chemical staining than lacquer or enamel. The downside to the urethane paints is the isocyanate catalyst which is extremely toxic. It might be easier in this case to let someone else paint the cylinders and cases.

Note: Harley-Davidson makes a wrinkle-black paint designed for engines. Not only is it meant for painting motors, it can be applied in one application without the need to prime the parts first. This is especially handy if you are painting an engine that's already assembled or in the bike.

Powder paint

Think of it as pulverized paint attracted to the parts by a voltage differential between the paint and the parts to be painted. Once the parts are covered with the "powdered paint" they are slid into an oven where the high temperatures bond the paint to the metal.

Popular with motorcycle builders, this process works very well for some engine components and frames. Powder coated parts will resist rock chips and chemical stains like nothing you've ever seen.

The disadvantage of this durable paint material is the heat used to bake the paint on, because it can ruin any seals or non-metallic parts. Worse, the heat can change the heat treating of the metal and the size of the parts. The heat may also force oil out of the pores of the part - which will then ruin the powder coating job. Once again, the parts have to be really, really clean before you start. Note: some powder coaters prefer the

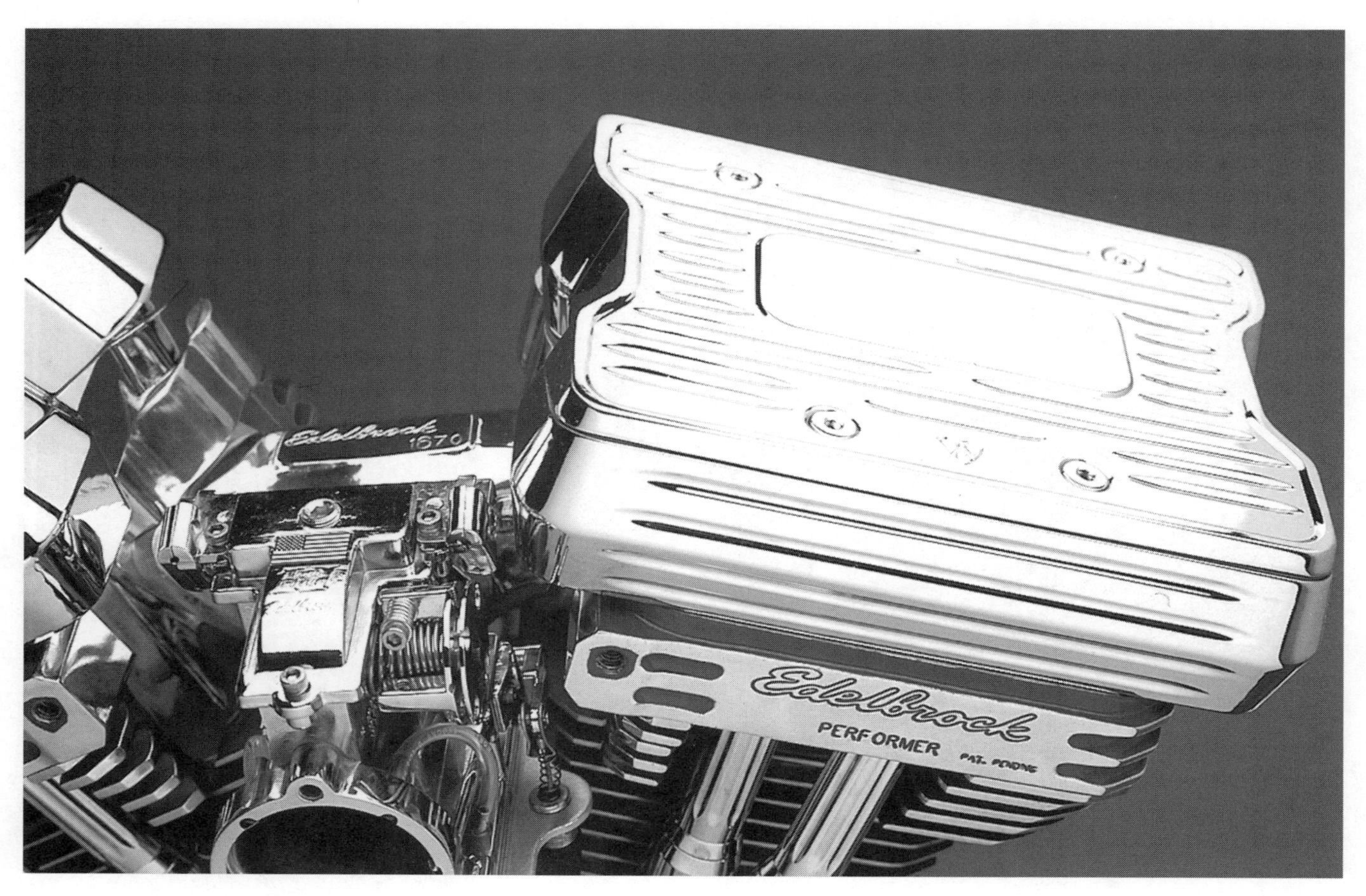

These two-part rocker boxes are chrome plated to provide the shine without the polishing maintenance.

Arlen Ness, Inc.

parts be bead blasted before the powder coating starts. If you bead blast the parts be sure *all* the residue is cleaned out of the engine parts before assembly.

Before powder coating cases and cylinders be sure to ask the manufacturer if the heat will damage the parts. If you decide to go ahead with the powder coating, have it done before the final machining so there will be no change in size or fitment of the parts.

If you're considering powder coating those engine parts, find a good operation in your area and then ask the operator for recommendations as to how the parts should be prepared before powder coating. Ask too about the color choices, some shops can provide everything from black to candy apple red, though they can't always match the exact color of a liquid paint system.

Chrome or polish

When it comes to making the parts shiny you've got two basic options, chrome plate or polish. Most of the external aluminum parts on your V-Twin can be either polished or plated. It's a matter of money, in that chrome plating costs more, which shine you like better, and how much maintenance you care to do.

Some people like the softer look of polished aluminum and don't mind the regular sessions with mild aluminum polish and a rag that are needed to keep that shine in top condition. Others prefer the look of chrome, or would rather spend Sunday morning riding than cleaning. This second group will go out of their way to buy parts, like rocker boxes, primary covers and so on that are already chrome plated at the time of purchase.

In the end you need to make sure the engine looks like it belongs in the bike. How you do it is a matter of your taste and budget. Remember that polishing, plating, blasting and painting all leave abrasives or contaminants on the metal that are detrimental to the life of your engine. So no matter what you do to the outside of the engine, make sure it doesn't hurt the inside of that same engine.

You can still have a great looking engine without spending all those bucks on billet. Use paint, especially on the air cleaner, like painter Jerry Scherer did on this Donnie Smith bike.

Chapter Eight

Exotica

When a little too much still isn't enough

How To Get It

I've tried to make this a book about building V-Twins for most of the people most of the time. The emphasis up to this point has been on carbureted engines of less than 100 cubic inches. This chapter is intended to answer the questions of people who just have to have more. More power, torque, maintenance and in some cases, complexity.

Extra power, over and above the abilities of a high performance 80 to 100 cubic inch V-Twin, can be had

The Big DUX from Hyperformance - a perfect marriage of German engineering and Yankee ingnenuity resulting in a 144 cubic inch V-Twin.

in three basic ways. First and most common, you can add cubic inches. Less common, you can make that 80 cubic inch engine think it's a 120 inch stroker by force feeding it air from either a turbocharger or supercharger. Last though not least, you can learn to laugh at the competition by placing your engine "on the bottle."

Cubes And More Cubes

If 120 horses is what you want you can achieve that power level with an 80 cubic inch engine, but that "little" engine needs to work pretty hard to obtain the power. Either it's equipped with just the right combination of cam, carb, heads and exhaust, or it's equipped with a blower, turbo or NOS. The other way to obtain the same horsepower level is with a larger motor - of say, 120 cubic inches - which will provide the required power output with a generally broader torque curve.

If you decide to belly up to the bar for a double shot of cubic inches you need to be sure the heads will match up to the cylinders and the cylinders to the cases. These major engine components all work together (at least they share the same cylinder stud pattern) up to a bore size of 3-13/16 inches. When the bore size goes to 4 inches or more, the cylinder stud pattern changes. Worse, some engines don't even use studs, but instead bolt the cylinders to the cases and the heads to the cylinders.

Because many of these cases move the lifter bores outboard you will probably need a longer pinion shaft, but even the extra long pinion shafts aren't all the same length.

The good news is the surge of popularity these monster motors are currently enjoying. This means more companies are manufacturing quality parts so there are more components and kits to chose from. The other added bonus is the relatively large number of manufacturers and shops who will sell you a complete monster motor either in kit form or already assembled.

The other "worry" in buying from the Big & Tall store is the frame fit. Most cases, with the exception of the really exotic billet cases, are designed to bolt up to the standard mounts in a standard frame, but that doesn't mean the frame is tall enough. We all think of strokers as being the tallest of the motors, but some builders of big-bore motors like to use longer connecting rods and some of the billet heads are a good bit taller than a cast cylinder head. All of which means that you'd better check to ensure the intended motor will fit the intended frame *before* laying down your hard earned cash.

The biggest downside to buying extra cubes is the extra expense of buying low-volume parts and the care that must be taken to ensure that all the parts will work together. The obvious advantage is the massive

This 116 cubic inch V-Twin is a combination of STD cases, S&S 4-5/8 inch flywheels, 4 inch Big Jugs and STD heads. Hyperformance

Hyperformnce makes their ductile iron Big Jugs in various sizes and configurations for everything from drag race motors to street applicatons. Hyperformance.

amounts of torque and horsepower available from massive displacement.

Extra cubes may not have the eye appeal of a blower or turbo, but they provide plenty of low-stress power with minimal maintenance and mechanical complexity. Big inch engines can be left in a mild state of tune so they idle and run without protest in slow traffic. Yet these same mild engines make tremendous power simply because of their size.

Blowers And Turbos

Feed more air, with matching amounts of fuel, to an internal combustion engine and it will produce more power. If you can't get enough air in those cylinders with big cams and ported heads add a pump and force-feed more air into each cylinder on each intake stroke.

Air pumps for engines come in two varieties, belt-driven superchargers and exhaust-driven turbochargers. Either device can add significantly to an engine's output.

The mechanical complexity offered by a typical blower installation can add a certain "wow" factor to the right bike. A polished blower housing with a chrome plated carburetor, or two, bolted on top is hard to beat. Though the blower, belts, and pulleys add complexity to the bike that complexity can add to the visual appeal of the motorcycle.

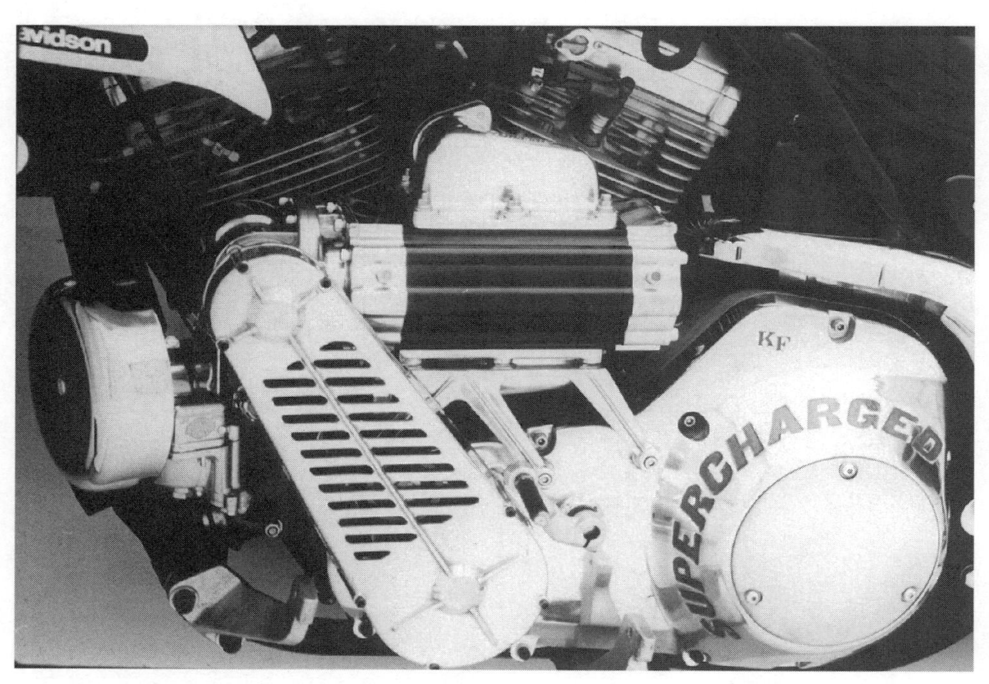

This is one of the compete kits available from Brilhante Company, installed on a factory FXR. More information on their kits and supercharging in general is available by ordering their What Is Supercharging book. Brilhante

Harnessing the engine's power to turn the blower generally requires at least one belt and two pulleys. Blowers have been mounted on both the left and right side of V-twin engines. The left side location puts the blower closer to the drive takeoff point, which is good. But now the output from the blower must either enter from the left side, remember the ports are angled out to the right, or be plumbed around to the right side.

Placing the blower on the right means that first the power to drive the blower must be routed from the primary side to the right side. Once the power is run across the bike with a short shaft you

This Donnie Smith creation mounts the blower on the right side, necessitating the drive and cross shaft which takes power to the K.F. Supercharger.

No matter how you drive the pump the air is heated as it is compressed. A given volume of air contains a certain amount of heat. When you compress that volume of air you have the same amount of heat in a smaller volume of air. Some turbo kits include an intercooler, or air-to-air heat exchanger, to reduce the temperature of the charge after it is compressed and before it enters the engine.

If turbochargers have the advantage of being less complex they suffer from a phenomena called "turbo lag." When you open the throttle,

still need a 90 degree drive to connect the power to the blower.

Blowers mounted down low in front, about where an air dam would mount, are easy to drive but require complex plumbing to get the air fuel mix to the blower and then back to the engine. Long intake tubes create a challenging tuning situation trying to allow for the cooling effect of the long tubes on the intake charge.

Less complex from a mechanical perspective is the option of turbocharging. Instead of driving the pump with belts and pulleys, why not use the energy contained in the exhaust as it makes it's way toward the open atmosphere. Essentially a turbocharger mounts the driven turbine in the exhaust flow. On the same shaft is mounted the drive turbine, or compressor wheel. Think of it as an exhaust driven blower.

Because the exhaust energy is normally wasted the power used to drive a turbo is often referred to as "free" energy. Yet, the fact remains, there is no free lunch. Turbocharger installations add complexity and expense to any engine and the presence of the turbo impeller in the exhaust affects the performance of the engine even when it is off boost.

Though it's a matter of taste, turbos are less "mechanical" than blowers and don't have the same aesthetic appeal. And because the turbo uses exhaust energy to run they affect the sound of the motorcycle.

The bower drive comes across the bike to the Don Hotop built 90 degree adapter bolted to the front of the polished blower.

An Engine Named Keck

If a 96 cubic inch V-Twin just won't cut it for your new project bike, then consider the Keck. Developed by Ray Keck, longtime motorcycle enthusiast and owner of an industrial manufacturing facility, this 4-1/4 by 4-1/4 inch billet engine is designed to take over where smaller, cast engines leave off.

The Keck is different in nearly every way, from the substantial size to the way it is manufactured and the material it is manufactured from. The raw material here is billet aluminum carved in-house to very precise tolerances on a CNC machining center. Both the cases and cylinders are cut from an alloy that the Keck folks refuse to identify. No, it's not the more common 6061 T6 or even the aircraft-grade 7075. The unidentified alloy is said to be stronger than either of the more common alloys - without any heat treating. Compared to those other alloys, this mystery material looses less strength as it is heated and will not anneal (which means going back to its pre-heat-treated softness) after thousands of heat and cool cycles.

The Keck engine moves the lifter bores outward to make room for the large diamter cylinders. This means you need a longer pinion shaft but can use all standard camshafts and valve gear.

This is a motor that was designed for cubic inches from the very start. Not only is everything heavier and more robust than more standard cases and cylinders, the lifter

This is a complete Keck engine with cylinders, ready for assembly at Huey Schwebs' Performance Engineering shop in Cleveland, OH.

bores have been moved over 3/8 inch to accommodate the larger bore without compromising the design of the cases. A longer pinion shaft, available from Jim's, can be mated to the flywheel assembly with ease and the offset lifter bores do not affect the engine's ability to utilize standard V-Twin camshafts and valve gear.

Currently there are no "Keck" heads, though both Patrick Racing and Hyperformance make heads that work just fine on the Keck engine. Forged pistons, sized to fit the ductile iron liners in the billet cylinders, can be ordered for your new engine, or a variety of aftermarket pistons can be used as well.

Keck Engineering reports rear-wheel horsepower figures ranging from 100 to 150 for "gas" engines running carburetors all he way to 200 and more for Kecks on laughing gas.

Keck reports that, "A standard Harley-Davidson flywheel assembly will work with the longer pinion shaft, but S&S makes a set of wheels that are easier to balance. We use a Carrillo eight inch rod which is super strong and gives us a better rod-to-stroke ratio than the smaller V-Twins. An appealing characteristic of this motor is its smoothness. It idles like a dream and enjoys a seemingly endless power band while accelerating." Though the Keck manufacturing facility is located in Ohio they have dealers set up throughout the country. Engines can be purchased raw, to be assembled by your favorite engine builder; or in various states of completion.

The Keck engine with matching billet heads from Hyperformance has an unusual "machined" look unique to the billet design and very different from cast engines.

RB Racing's turbo kits come complete with turbo, wastegate, exhaust, oil lines and hardware, air cleaner and intercooler. Fuel injection is optional. RB Racing

asking for more air, it takes time for the turbo to "spool up" to speed and provide the necessary boost. Some of the newer systems use adjustable vanes on the driven wheel to help the turbo adjust to different conditions and reduce turbo lag.

Because the turbo wheels spin at incredible speeds and the shaft bearings are subject to heat from the exhaust, lubrication of the bearings is a key issue. While some systems plumb oil to and from the bearing housing, certain new aftermarket turbos (and blowers) simply use a reservoir of oil to lubricate the bearings. Clever designers have gone so far as to locate the bearings closer to the intake side of the turbo housing so the bearings are cooled by incoming air instead of being heated by the exhaust.

Regardless of which style of pump you use a "stock" engine can only accept so much boost before increased power, pressure, and heat causes it to self-destruct. Most kits designed for the street keep boost in the six to eight psi range. Turbos have more potential for runaway pressure and commonly use a waste gate or control mechanism to limit boost.

RB Racing puts their IHI turbo out front where it stays cool and doesn't overheat the rider. Each kit comes with a boost gauge and Dial-A-Boost adjustment. RB Racing.

If you like the idea of forced induction call around for more information before you start buying the parts for that new engine. Though most of these kits are designed to be bolted onto stock bikes you can get the most from the installation if you consider the blower or turbo at the very start of your engine building process. Specifically, bower and turbo installations work better on engines with lower compression and less valve overlap (a wider lobe separation angle). Many of these turbo and blower kits come in "street" version or a "competition" model with boost levels that reach 25 psi and more.

Where To Buy Blowers

There aren't just a whole hell of a lot of companies making blowers sized or designed for a typical V-Twin. The short list might start with the neat "billet" blowers seen on some high-tech show bikes and customs. These nifty little blowers were manufactured by K.F. Engineering. Today the same unit, available in different sizes, is manufactured by Fageol Superchargers. These are roots-style blowers with steel rotors in an aluminum center section. Four different size superchargers are available as well as a series of different drive assemblies, including a ninety-degree drive.

Brilhante Company in Los Angeles California makes a blower kit for V-Twins which mounts neatly on the engine's left side. This kit, which uses a blower from Fageol, is available in two models: The "Max 15" will produce up to 12 psi boost (boost is determined primarily by the drive ratio) while the "Max 30" can huff and puff all the way to 25 psi.

Camden Superchargers from Austin, Texas makes a compact blower based on an aluminum housing and aluminum rotors with Teflon seals. This compact blower, available in various sizes, uses pressure lubrication from the engine to provide long bearing life and cool running temperatures. Camden Superchargers are used by R.B. Racing in their V-Twin kits.

Whipple Industries in Fresno, California makes a twin screw style of blower, available in various sizes, which is said to offer significant advantages for the more common roots-style blower. Specifically the twin-screw design is said to produce a greater volume of air at a cooler temperature than other blower designs. More information on Whipple charges is available

Aerocharger kits include everything you need for a complete, high-horsepower installation, including the intercooler and carburetor. Nempco

from Advanced Racing Technology in Burgess, Virginia.

TURBOS

Garrett Turbochargers (originally known as AIResearch) come in a number of sizes and are used in a variety of vehicles including automobiles. These turbochargers are available with different style housings and adjustable waste gates. They even offer a special variable boost kit known as dial-a-boost. Garret turbochargers are sold and distributed by Turbonetics, a company with a wealth of experience in turbo installations.

Mr. Turbo from Houston, Texas offers a draw-through kit - as opposed to those that "blow through" the carburetor - based on a Garrett turbocharger with integral waste gate. By adjusting the waste gate maximum boost can be set anywhere from five to fifteen pounds. Though five pounds of boost might only get you eighty some horses, depending on engine size etc., fifteen pounds will produce 130 horsepower from an eighty cubic inch V-Twin.

The Mr. Turbo kit includes the turbo itself, a S&S carb, the necessary manifold, a chrome plated exhaust system, and a nacelle that hides the turbo.

Aerodyne makes a range of turbos with at least one ideally sized for V-Twins. With its own lubrication system and variable vanes to help the turbo spool up faster at almost any rpm, this turbo offers a wealth of advantages for motorcycle work. The Aerocharger kits - offered by Nempco dealers as well as First Choice Turbo Center - use an Aerodyne turbo mated to all the necessary intake and exhaust plumbing (including an intercooler) to create a very neat installation capable of producing 125 horsepower.

I.H.I. is another company who manufactures turbos for a wide range of applications. One of their compact, light weight turbos was used in the first-generation V-Twin turbo systems from R B Racing. Currently RB Racing has a number of new turbo systems available for engines ranging in size from 80 to 160 cubic inches. R.B.'s "street" systems can take your V-Twin to 275 Horsepower (guaranteed to eat Ninjas for breakfast) with 15 psi boost and fuel supplied by their own very sophisticated fuel injection system.

Note: If you want more information on blowers, turbochargers and nitrous try *Turbo Bike and High Performance Magazine.* This is not another standard "motorcycle" magazine with technical articles written by the advertisers, illustrated with the help of silicone-enhanced models. *Turbo Bike* is seriously technical, with coverage of Japanese four-bangers as well as American V-Twins, and incudes a wide variety of articles about high performance motorcycles both on the street and the strip. Contact Joe Haile Ent. PO Box 800725, Valencia, CA 91380.

You can have your nitrous the easy way, with a compete kit from Nitrous Oxide Systems Inc. Each kit contains a bottle (available in diferent sizes, nittous and fuel solenoids, fogger nozzles, filters, fittings and all hardware. NOS

Randy at Hyperformance keeps a number of connecting rods in stock, including the forged S&S designs and the lighter offerings from companies like Carillo.

NITROUS OXIDE

We've all heard about nitrous oxide, variously known as laughing gas, or "squeeze." It makes fast bikes even faster, and trips to the dentist bearable. Just hit the button and the front wheel is guaranteed to reach for the sky. Despite all the things we think we know many among us don't really understand how this stuff works.

A nitrous oxide molecule is made up of two atoms of nitrogen and one of oxygen. What it brings to the party is not more fuel, but more oxygen - so you can burn more fuel. NOS exists as a liquid when it's under pressure in the small tank. This liquid changes to a gas as it passes through the jet and into the intake manifold. This change of state absorbs a tremendous amount of heat which helps to create a nice dense intake charge.

Nitrous has the added advantage of "power on demand." You can run nitrous on nearly any engine, from mild to wild, but it only adds to the power output when you're "on the button."

Typical power increases in the thirty to forty percent range are common on V-Twins with a kit from a company like NOS in Cypress, California. The complete kit includes the bottle, available in different sizes, nitrous and fuel solenoids, fogger nozzles, filters, fittings, a fuel pump and all rest of the hardware you need for installation.

Most nitrous systems are designed to operate at full throttle only, and often only in the upper gears as nitrous used at low rpm can damage the engine. Essentially, when you hit the button both the nitrous and the fuel solenoids open, adding oxygen and extra fuel (to go along with the extra oxygen) to the engine. The result is a burst of power sure to catapult you ahead of that damned rice-rocket in the other lane.

For truly jaded horsepower junkies nitrous can be combined with a supercharger or turbo installation for awesome power levels. In these situations the cooling effects of nitrous are a nice added bonus.

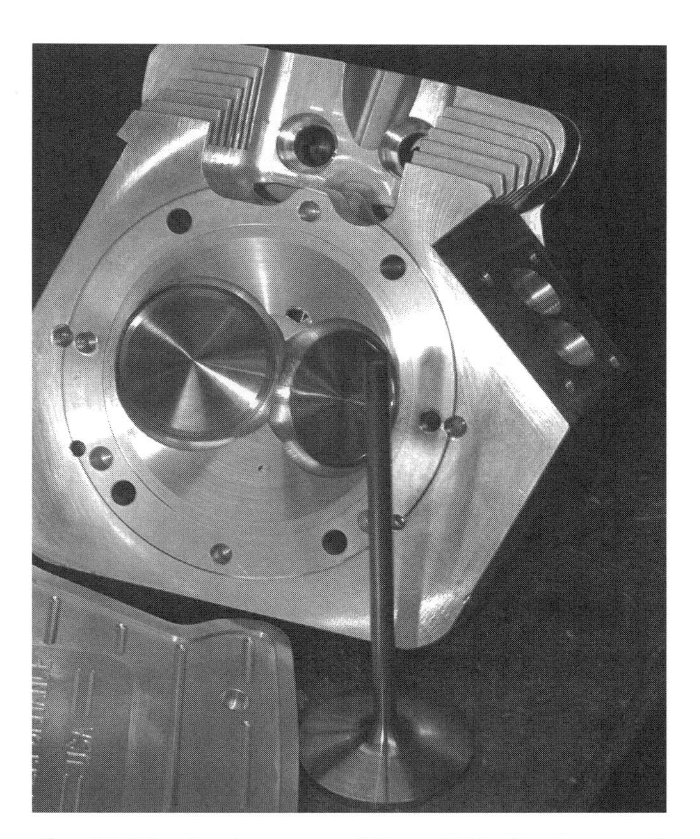

Randy's Megaheads are carved from 6061 aluminum and designed to fit engines with bore sizes up to 5-1/8 inches. Valves measure 2.6 and 2.0 inches for the intake and exhaust respectively.

Shop Tour

Hyperformance, Des Moines, Iowa.

Randy Torgeson might be called the King of Cubic Inches. Though he started Hyperformance in 1983 doing mostly cylinder head modifications and restoration work (and still does a fair amount of that work) he has developed a reputation as the man who can put together a V-Twin that measures up to 179 cubic inches.

The "entry level" engine at Randy's Cubic Emporium is a 116 cubic inch engine based on cast cases from STD mated to his own Big Jugs and STD heads. The Hyperformance Big Jugs are bored to a four inch diameter, matched to a 4-5/8 inch stroker flywheel assembly from S&S. This particular V-Twin is available in a HyTorque version with heavy flywheels, the flywheels come from either S&S or Truett & Osborn, and round-port heads from STD. The HyHorse rendition of the same motor comes with lighter flywheels and oval-port heads for a motor with quicker revs and faster acceleration.

If you want beauty with your beast Randy can fix you up with a complete, billet Keck engine (see sidebar in this chapter) topped off with his own billet Mega Heads.

If you're thinking of a really big engine, but one that you can still drop into a factory frame without any hassle, then consider the 116 or 120 cubic inch engines. Randy reports that cast versions of either of these two monsters will fit into a standard frame and bolt up to a stock inner primary. As you add cubic inches and/or billet engine cases however, the engines grow externally which makes them a tight fit (to say the least) in most frames. Many aftermarket frames are taller than a similar piece from the factory, but each one is different. If you're building a bike from scratch and you intend to go for maximum cubic inches consider the height of your new engine before you buy that new frame.

Though some of the new monster motors are designed to use aluminum cylinders with liners Randy manufactures his own Big Jugs from billets of ductile iron. If you ask him why he prefers the ductile iron he replies without hesitation, "It doesn't grow as it heats, the coefficient of expansion is virtually zero. And ductile iron is a very strong material with a tensile strength of 90,000 psi compared to 35,000 for cast iron."

"We design and build our cylinders with a wide foot print for a very stable cylinder. It helps to counteract the way V-Twins are always trying to force themselves apart. By bolting the cylinder base to the case and then bolting the heads to the cylinder we create a much stronger design than we would by using through-bolts. This way the heads screw into ductile iron instead of into aluminum."

Randy uses Ross forged aluminum racing pistons for most of these monster applications. Rather than a barrel-shape these Ross pistons have a .003 inch taper as measured from top to bottom. Fit for the Ross pistons is on the loose side - .008 or .009 inches as measured at the skirt. The exception is the 5.1 inch Big Jugs which use barrel-faced JE forged pistons - unless of course the engine is set up to run nitro!

Connecting these huge pistons to the crankpin are a some very high quality, and long, connecting rods. "We stock rods all the way up to ten inches in length," explains Randy. "Some of those are titanium to help keep the weight down. For the actual length I like a

Randy Torgeson is the power behind Hyperformance, the man responsible for motors like the Big DUX he's leaning on.

rod length-to-stroke ratio on the order of two-to-one"

If 120 cubes isn't enough, then you can order one very exclusive piece, a 156 cubic inch engine based on billet cases from DUX, a German company who originally came to Randy for help developing the cases for their "DUX big block." This over-square design uses Randy's Big Jugs with an internal dimension of 5.1 inches and a flywheel assembly with a stroke of 3.813 inches. On top is a pair of Hyperformance Mega Heads with a 2.6 inch intake and 2.0 inch exhaust valves.

The current "king of the hill" at Hyperformance is a 179 cubic inch V-Twin based on Morrocco billet cases equipped with a S&S flywheel assembly with a 5.375 stroke, Big Jugs with a 4.6 inch bore, 9 inch HyTech connecting rods and a pair of Mega Heads.

Randy stocks a variety of pistons in large sizes, including these examples which have been modified to provide piston-to-piston clearance at BDC.

Randy helped the German company, DUX, develop the crankcases for this 144 cubic inch V-Twin. Cylinders and heads are both Randy's designs. Hyperformance.

When asked about the problems people encounter when they go this route, Randy talked again about making sure the motor fits the frame, "That billet DUX engine is so big you've almost got to build a frame around it. And sometimes people get more than they bargained for. Some of these big engines are just damned scary. I prefer to build them for torque and then tell people to gear them as high as possible, because you can't bog one of these things."

Are they hard to start? Yes, but it's not an insurmountable problem. Randy recommends first that owners install a really good battery before they go out and buy the high-torque starter. "We've had good luck with the Yuasa YB16HL-A-CX, it's got more cranking power in the same size case as some other batteries. There are also some neat little compression releases that can be screwed into the cylinder head, with a hole tapped for them next to the plug hole, and then opened during difficult cranking situations."

So if way too much still isn't enough there are some engines out there just for you. They don't come cheap, but they do run and they are available. What you might call true hy-performance.

Chapter Nine

In The Shop

Assembly sequences

WHAT THIS IS - AND ISN'T

The chapter includes a variety of assembly sequences. These are intended to help readers assemble their own engines and to become more familiar with the innards of a modern V-Twin. As I've said before, these sequences and this book do not take the place of a good service manual. Also, the bottom-end of a V-twin should be assembled by a trained mechanic with all the necessary special tools.

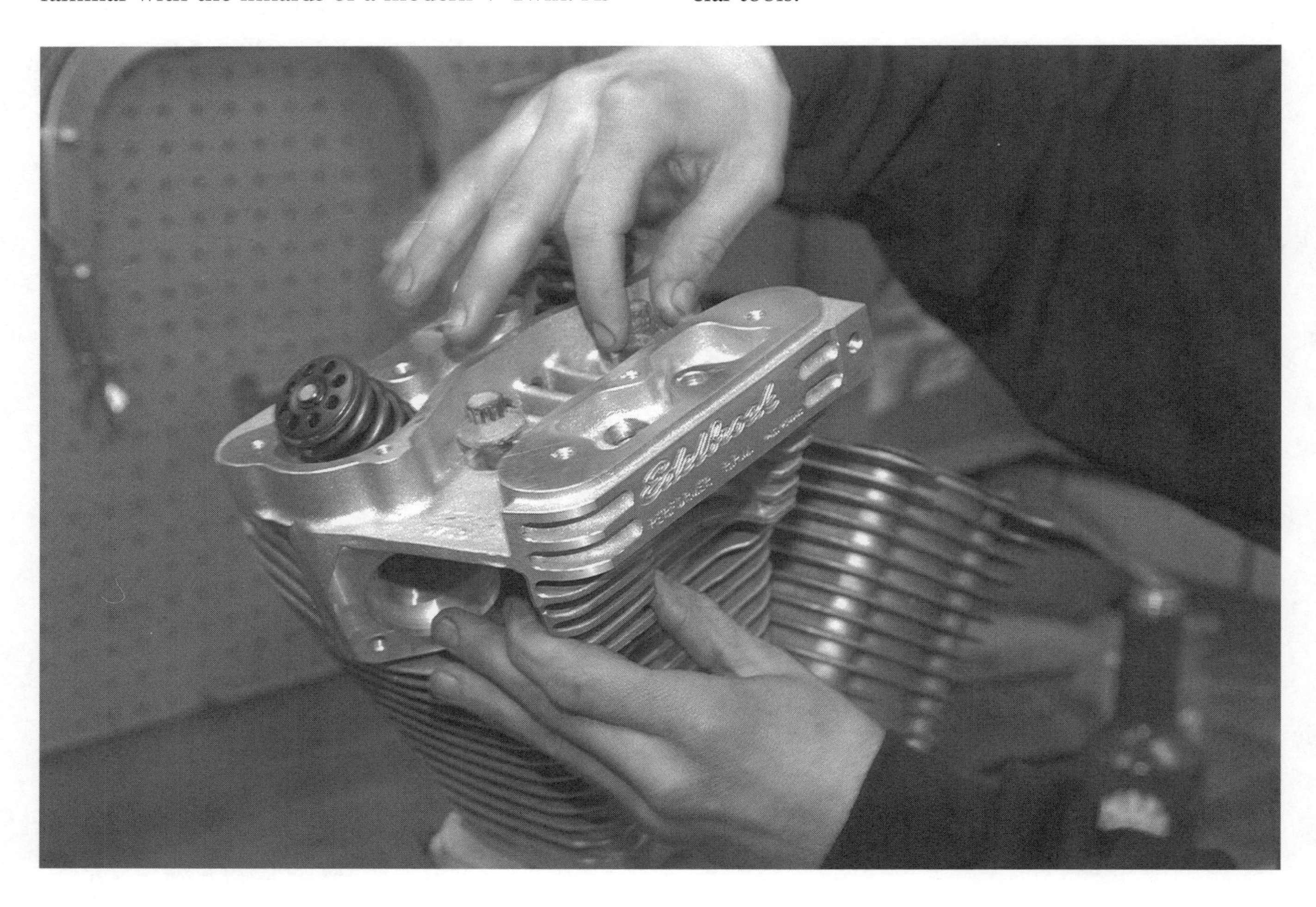

The hands belong to Steve Tuttle from John Bryant's M-C Engineering in Cleveland, OH. Steve lubricates the head bolts before installation for more accurate torque wrench reading.

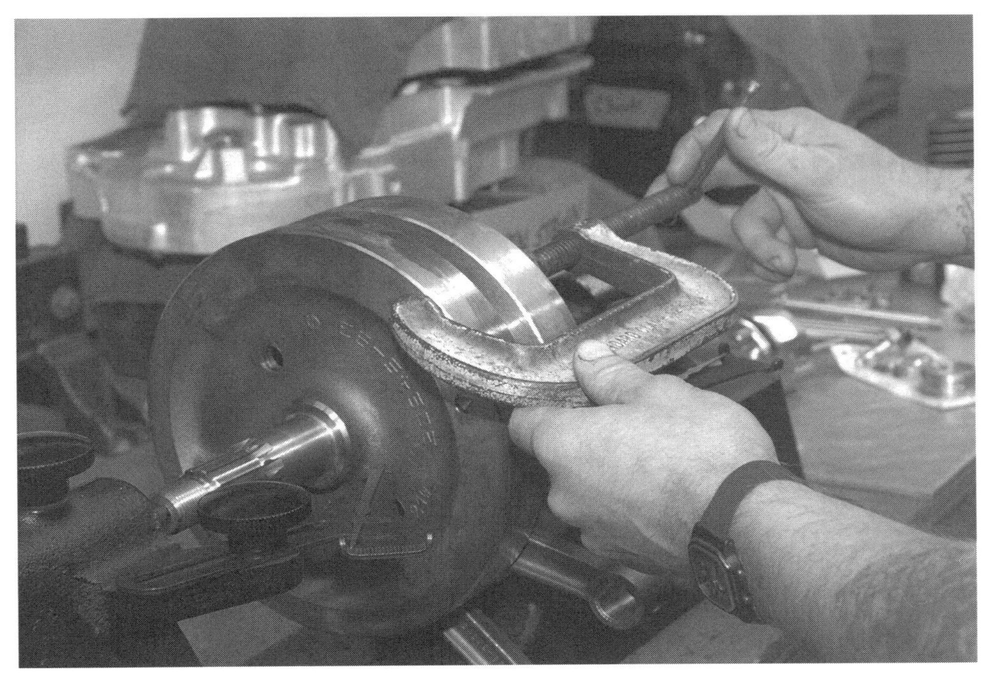

Runout can be corrected by spreading the wheels, or pinching them with a clamp like that shown here.

fixture and checks the connecting rod end play in the flywheels. The check is done with a feeler gauge. The specification is .005 to .025 inch. This set is found to be .013 inch.

Once the flywheel assemblies are known to be true and the wheel-to-connecting rod end play is within specification the assembly of the bottom end can proceed. The left side of the flywheel assembly is supported by two, tapered "Timken" bearings. The first step is the installation

Though the flywheel assembly is new, Lee Wickstrom likes to check the runout in the fixture seen here.

As a final disclaimer, these procedures are for a "stock" motor. Lee from Kokesh MC made the following comments: Stroker engines should have the piston-to-piston, rod-to-crankcase and cylinder spigot, and piston-to-flywheel clearances checked. You may also want to "blueprint" the breather timing and "degree' the camshaft.

Assemble The Bottom End At Kokesh Mc Parts

The flywheels

We start this sequence with a set of new 80 cubic inch flywheels from Harley-Davidson. Lee Wickstrom, the engine man at Kokesh, starts the building process by checking the flywheels to make sure they are true. Lee explains that, "The maximum allowable runout is .002 inch, but I don't like to let them go past .001 inch. Typically they are very close as delivered, but you want to make sure they haven't been bumped in shipping"

To correct the wheels Lee would pinch or spread the flywheel assembly. Shown in one of the photos is the clamp Lee uses to pull the two flywheels together and correct a high spot.

Next, Lee mounts the flywheel assembly in a

A feeler gauge is used to check connecting rod end play between the wheels.

Here the installation tool is used to install the inner Timken bearing.

The same tool used is used on the outer bearing. Here you see a dial indicator mounted to check the end play.

of the inner bearing, which Lee does with the help of a special bearing installation tool. He warns that anyone who installs the bearing with a piece of pipe and a hammer will hammer the wheels out of true as they pound the bearing onto the shaft.

Before going on with the assembly sequence I should back up to explain that the two tapered Timkens must be installed so the shaft has just a very small amount of end-play. The factory specification is .001 to .005 inches. Lee likes to set the bearings up so the end-play is slightly less than .001 inch - but there is no drag or pre-load on the bearings. The end play is determined by the spacer used between the inner and outer Timken bearings. Lee notes that S&S cases come with the races for the Timken

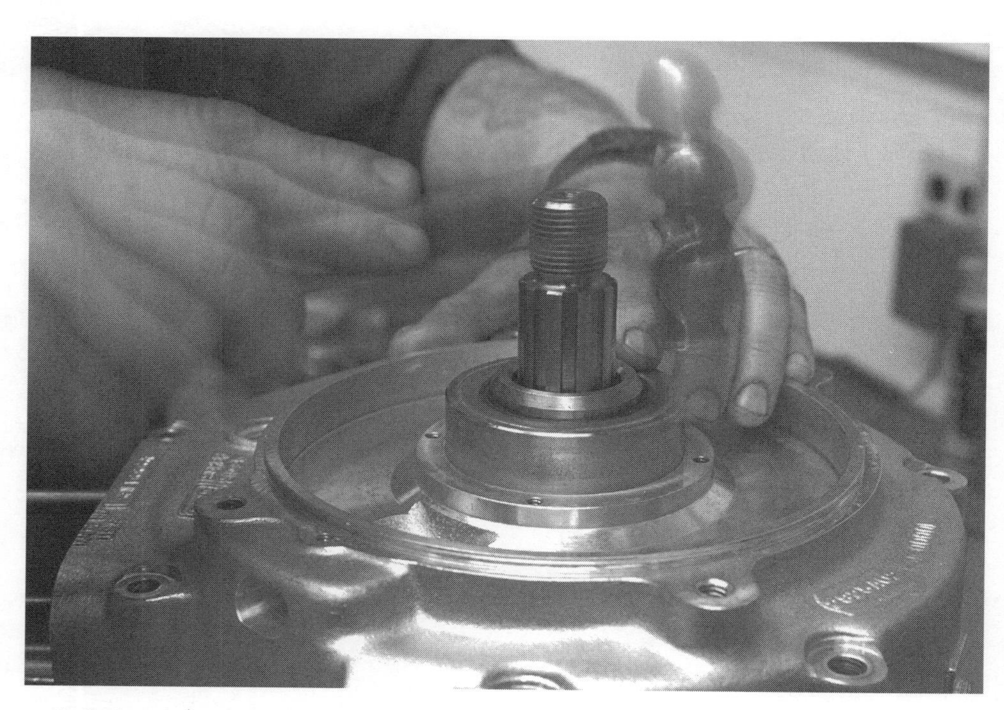

A light hammer is used to install the sprocket shaft seal per the instructions in the service manual.

bearings already installed. The cases come with the Timken bearing set: two bearings, two races installed, and inner and outer spacers.

Lee notes also that in most cases where the Timken bearings are already installed, if you find there is no end play and a little drag in the bearings it may be that the Timken races are not seated tightly. These races must be pressed together, installing them with a hammer allows them to "bounce" resulting in a race that isn't seated tight in the case.

Installation lube and a driver are used to minimize the force needed to install the outer camshaft bushing. This bushing will be reamed to size after installation.

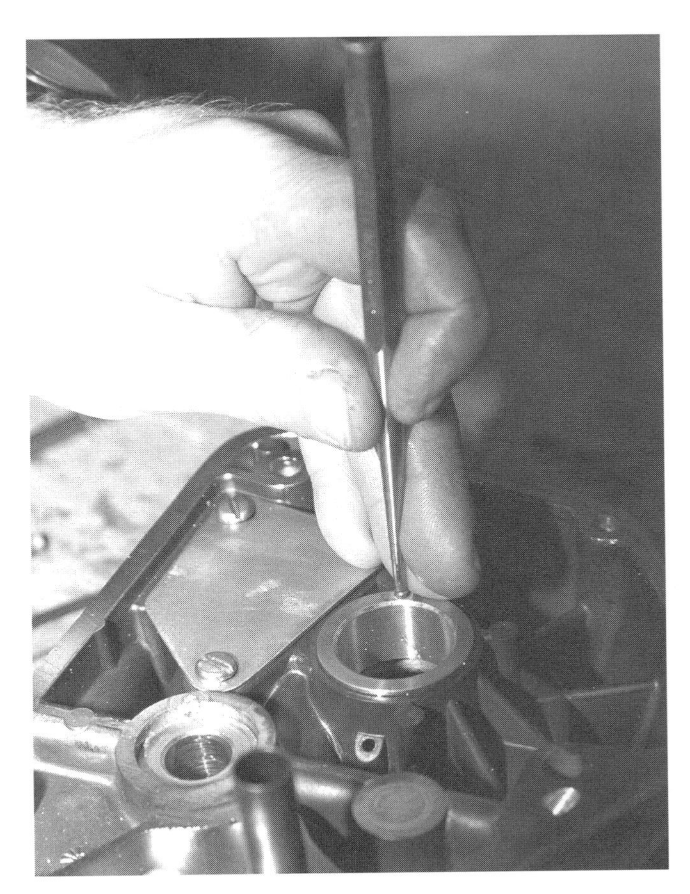
After installing a new lock pin, Lee uses a small punch to peen material over the pin so it can't back out.

The cam bushing reamer passes through the inner cam "Torrington" bearing. After the cam cover is bolted on the reamer is used to size and align the bushing.

After reaming the cam bushing to size and removing the cam cover (with the correct puller) Lee drills the oil hole in the bushing.

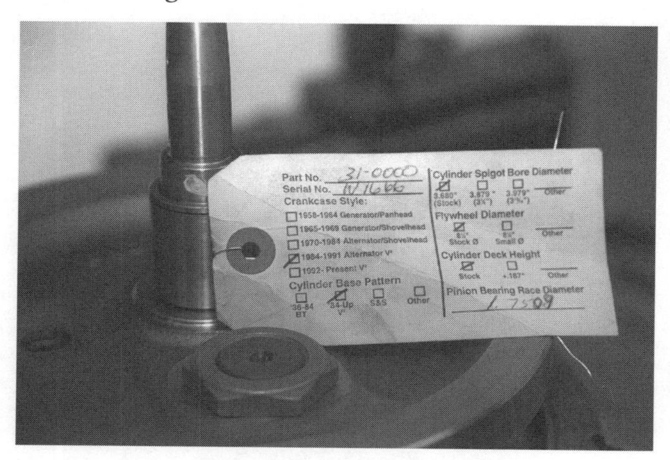

Here is the tag which came with the new S&S cases, and the color coded paint on the inner crank bearing race. A chart in the service manual will determine which pinion shaft bearing will match up to the inner race O.D. and the outer race I.D. to provide the correct clearance.

These pinion shaft bearings are available in various sizes as mentioned in the previous caption.

Install cases

Back to our flywheel installation. With the flywheel assembly in a fixture Lee sets the left case half, with the bearing races installed, onto the flywheel and then installs the inner Timken spacer (which determines the end play of the flywheel assembly) and then the outer Timken bearing with the same special tool used for the inner bearing. Both bearings are coated with assembly lube before installation.

Next Lee installs the sprocket shaft seal and seal spacer. There is some confusion as to which way the lip of the seal should face. "The direction that the seal faces depends on whether or not you are running a wet or a dry clutch," explains Lee. "Consult your service manual for current recommendations."

Before the right side of the engine cases are installed Lee first prepares the cam cover by installing a new outer cam bushing. The bushing installs with an interference fit so he uses installation lube to ease the installation. Next he drills a hole for the lock pin with a number 31 drill bit. The pin is installed so it's under-flush, then Lee peens over the top of the bushing so the pin can't possibly back out. The new cam bushing comes with no hole for oiling so after the installation is complete Lee drills the oil hole in the bushing.

Note, the other end of the cam is supported by a roller bearing. Up until the late 1980s the factory used a true Torrington bearing for this task, but later engines from the factory use a replacement bearing with fewer needles and thus less support for the cam. Lee will only use the Torrington bearing, available from most dealers or the aftermarket (Drag Specialties #DS-198805) for maximum support. Once the inner cam bearing is in place Lee can slip in the reamer - used to size and "line bore" the outer cam bushing - from the back side of the right side case. Then the cam cover is installed and the reamer is turned by hand. After pulling the cam cover off the case it's time to thoroughly clean all the metal shavings from the case and the cover.

Before installing the cam-side engine case, Lee checks the color code on the inner race installed on the crank, and then the size noted on the tag that came with the new S&S cases. Next he refers to the chart in the service manual which tells him which bearing to match with this particular combination of race I.D. and shaft O.D. Note, these measurements are done to the hundred-thousandths of an inch so it's got to be very, very accurate.

This particular example falls in a grey area where either of two bearings will work. In this case Lee can have the clearance looser or tighter, and makes the comment that, "Because this is Jason's motor (another Kokesh employee) and I know he's a good rider who's going to break it in right, I will set it up tighter."

Lee double checks the pinion shaft O.D. and the pinion race I.D. to be sure his measurements match the factory measurements and settles on .0004 inch clearance.

The last thing Lee does before installing the right side engine case is check for adequate clearance between the innermost cam lobe and the case. Some cases do not allow for the extra lift of high-performance camshafts, in which case the mechanic must use a die grinder or a special tool available from Zipper's to "clearance" the case (see the photographs for a better understanding of this potential problem).

Lee uses Gasgacinch on both case halves, lets it get tacky and puts the case halves together. Note, three of the studs are a snug fit through the cases and are used to align the cases, Lee installs these three first.

The case bolts must be torqued down in two stages, using 18 foot pounds as the final torque, following the tightening sequence in the service manual.

With the crankcases assembled it's time to check to be sure the rods are straight before proceeding with the rest of the assembly. Lee uses a "checking rod" from S&S. This is like an extra long wrist pin. Once installed in the connecting rod the engine is rotated to bring the pin down on the deck. As Lee explains the procedure, "with paper under each side (check the photos here) and the pin resting on the two pieces of paper, there should be the same amount of drag on each piece

Before bolting the cases together it's a good idea to check that there is enough clearance between the case and the inner-most cam lobe.

After installing the "locating" bolts first, Lee uses a torque wrench to tighten up the case bolts.

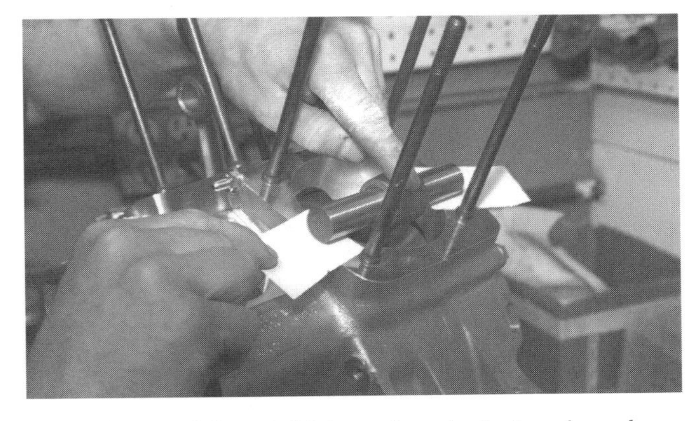

A checking rod from S&S is used to check that the rods are straight. The pressure on the paper under each side of the special rod should be the same.

It's a good idea to disassemble, inspect and clean the oil pump, even it it's brand new.

when you pull the paper out."

It is interesting to note that Lee keeps careful notes on the assembly of each engine. He finds this helpful in evaluating trends and wear patterns when he rebuilds the same motor years from now. It's also a useful "back up" in case of a part failure that might otherwise be attributed to improper installation.

The oil pump

First a few words about oil pumps for V-Twin engines. The pump is actually a double pump. The supply side takes fresh oil

The driven gear can be slid onto the oil pump shaft as the pump body is set in place. The drive key and snap ring will be installed after the pump is bolted in place.

The pump to case bolts are screwed in by hand, though the final tightening is done with a small torque wrench.

from the oil tank and feeds it to the engine. The separate, return side, pulls the oil out of the bottom of the cases, then through the pump, through the oil filter and back to the oil tank.

In this case Lee is working with a new oil pump. "I always clean out the oil pumps before I install them," explains Lee, "even if they're new. The S&S and Harley-Davidson are good pumps, but people should not chrome them because you get extra chrome built up at the edge, thus the pump might not seal and even if it does the clearances are wrong. I tell people to polish them or leave them alone."

After re-assembling the pump, Lee puts oil on the bolt threads. The driven gear must be put on the oil pump shaft as you slide the pump body up against the case, though the drive key can be installed later (check the photos to relieve confusion). Now Lee finishes bolting on the pump body and then installs the key. After the key is in place a snap ring pliers with 90 degree jaws is used to install the small snap ring that prevents the key from coming out of place. Note: the little snap ring must have good tension to stay in place, don't open it so far during installation that it looses its "spring."

Lee likes to turn the driven gear as he's torqueing the body down so he knows the pump is not binding and adds the comment, "The pump should turn nice and free even after everything is torqued in place.

Next he installs the oil pump drive gear and key, followed by a spacer and then another key and the pinion gear. A special socket is used for the pinion nut, Lee uses a torque wrench to tighten it to the proper specification.

Note: there are differences in the pinion shafts. Some are straight, some are tapered; some use one key for both the pinion gear and oil pump drive gear and others use two keys. Late model factory pinion shafts are straight and rely on the key to secure the gears to the shaft. Most mechanics feel the tapered pinion shaft is better than the later straight shaft as it provides better contact between the pinion gear and the shaft, and doesn't rely solely on the key for support.

Here you can see the oil pump drive gear (with key) set in place, and the spacer between the drive gear and the pinion gear.

The pinion shaft and key are slid in place, then tightened with a special socket and a torque wrench.

After installing the oil breather (plastic in this case) and an end washer, Lee checks the end play. Note that the new gasket is in place as Lee makes the check (he will compensate for the fact that the gasket will compress).

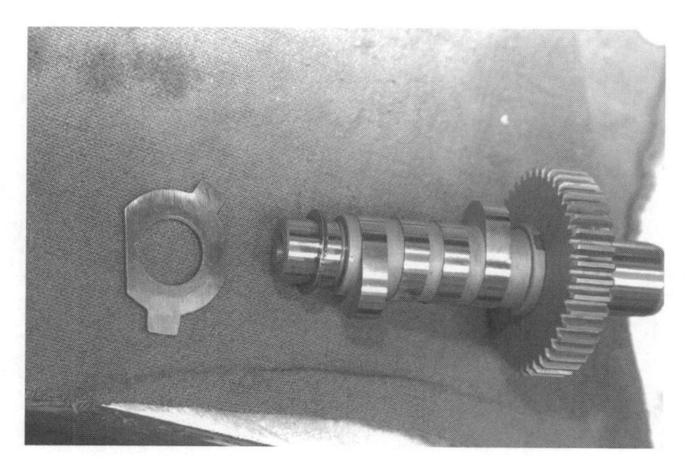

Here we see the cam with the factory thrust plate and the thrust washer (seen on the end of the shaft) which must be the correct thickness to provide the right end play.

Here the cam is installed. Before the final installation of the cam cover Lee will make sure the cam to pinion gear fitment is correct and that the cam has the right end play.

The final check for cam end play should be made with a dial indicator working against a bolt screwed into the end of the cam.

Before describing the installation of the breather gear we need to backup again and address the current controversy over this little gear assembly. The factory installs a plastic gear though steel models are available from various aftermarket sources. The popular press berates the plastic gear as being, well, plastic. The heavy-metal crowd is quick to point out the fact that plastic gears can pickup bits of metal which are floating around in the oil, allow them to become embedded in the plastic, which in turn creates a cutting tool inside your engine. Other experienced voices however, prefer the plastic because if something comes loose in the motor, and gets lodged inside the breather, the gear will split. If the same thing happens with a metal gear you run the risk of breaking the cases.

Regardless of whether the gear is plastic or steel it must be installed with just enough end play, in this case .001 to .015 inch as measured between the little end washer (which are available in various thicknesses) and a straight edge run across the edge of a new gasket. When measuring you have to remember that the gasket will compress about .005 inch. A little grease is used to hold the thrust washer in place until the cover is installed.

The Bumpstick

The cam must be installed with both the correct end-play and the right gear fitment. In terms of end play, it's a matter of using the correct thrust washer between the cam end and the thrust plate. The factory specification is .001 to .050 inches but most mechanics like it tighter rather than looser. Lee shoots for an installation with a total of about .006 inches of end play.

Note, there is a procedure for checking the size of a cam gear with special pins, as in the case where the old stock cam (which was fit at the factory to be nice and quiet) is being replaced with another camshaft. The pins can be used to measure the size of the cam gear and compare it to the original. But in the case of a new engine like this one you have to work pretty much by feel.

Lee likes to start the process by installing the camshaft and cam cover with no thrust washer on the cam shaft and no valve train in place. He explains that with the cam installed in this condition, "you should be able to reach in through the tappet block holes and move the cam back and forth. You should feel just a little drag caused by the gears sliding past each other. If you can't move the cam then it means the pinion and cam gear fitment is too tight. If there's no drag at all

The lifters, coated with assembly lube, are slid into the lifter blocks from the bottom. Use of high lift cams may require that the recesses for the rollers be enlarged.

then the gear fitment is too loose. Also, there should be just a little end play in the breather, as a double check on the thrust washer that was used on the end of the breather."

After checking the feel of the gear fitment and the end play of the camshaft Lee uses the correct puller to remove the cam cover, otherwise there's the risk of beating up the cover trying to get it off - this is especially true with chrome plated covers.

In this particular case Lee discovered that the cam gear to pinion fitment was tight at one spot in the rotation of the gears, causing him to carefully inspect and clean both gears. For camshaft end play Lee likes them fairly tight, "I like to set end play at about .006 inch. This is more important if you're dealing with a mechanical advance unit. I put in a small thrust washer, one of known dimension, and then check the end play with a feeler gauge after reinstalling the cam cover. For the final check though I like to use a dial indicator."

Because of the tightness in the fitment between the cam gear and pinion gear, Lee checks the pinion shaft runout and finds that it is .0025 inch. The specification from the service manual allows runout up to .004 inch so that is not causing the bind in the gears. The next step is to try a second camshaft, but the new cam exhibits the same tightness in the same spot so next Lee tries a new pinion gear.

Ideally there should be just a small amount of drag between the two gears when you try moving the cam back and forth without any thrust washer in place. The new pinion solved the fitment problem. Now that the fitment is correct there's still the camshaft end play to worry about. With a bolt screwed into the end of the cam Lee sets up a dial indicator to read directly off the bolt and finds that the end play is .013 inch. Lee will install a thicker thrust washer to bring the final end play to the ideal of .006 inches.

New Big Twins use a one piece thrust washer, the

Lifter blocks are assembled with "Tight" on the threads. Note the use of the tapered bolt to correctly locate the lifter blocks.

older style set up seen here uses a thrust washer of varying thickness that fits between a large diameter washer and the shoulder on the cam. With the new pinion gear and the correct thrust washer on the camshaft Lee installs the cam cover for the "last" time (though he will double check the end play just to be sure it's correct). Lee uses "Tight", a thread locking and sealing compound from Fel-Pro on the bolt threads (this is a good locking material for aluminum threads).

Lifter blocks

The lifter blocks are next. First Lee checks that the bores are within specification. In this case both the tappet blocks and lifters are new. The specification is .0008 to .002 inches of clearance, which Lee checks with a micrometer and the telescoping gauge.

After applying pre-lube Lee installs the lifters into the bores. then sets the assembled lifter blocks and lifters onto the engine. He uses a tapered bolt (a special tool) to correctly locate the lifter blocks on the case, then installs the other bolts using a little more Tight. The torque wrench is used for the final tightening.

Lee warns first-time builders that, "With a radical cam, over .560 inch of lift you should check that there is enough clearance on the bottom of the lifter between the roller and the slot in the lifter block, at maximum camshaft lift. There should be a minimum of .040 inches of clearance (some shops recommend .080 inches) and sometimes you have to create a deeper slot for the lifter roller.

Be sure it's oiling

Lee likes to prime the pump and make sure it's oiling correctly before installing the pistons and cylinders. His checking procedure for the oil pump goes like this:

1. Leave the check ball (see photo) out of oil pump until there is oil in that passage.

2. Now turn it over until oil starts to raise the pressure relief valve, which means you're moving oil into the motor.

3. Install the spring and cap on the pressure relief valve.

Note, the oil pump has three stages: the first stage feeds oil to lifters and top end. After more pressure

Here you can see the funnel Lee uses to feed oil to the pump during the thorough priming process he goes through. Punch shows location of ball Lee uses to force oil to the rods.

builds up, the second stage sends pressure to the lower end. The third stage is to relieve excess oil pressure as might occur during a cold start.

4. When priming and checking the pump Lee drops a steel ball down into the passage where the tappet screen usually resides (see where the punch is placed in the illustration) this blocks oil to the lifters so you can turn the engine over and be sure the rods are getting oil.

5. Then take out the extra ball, install tappet screen and continue rotating engine to see if oil goes to the lifters and lifter bores. "In this way we know that oil is going everywhere it's supposed to." explains Lee.

Top End Assembly Sequence At John Bryant's M-C Engineering

The next assembly sequence was done at John Bryant's Motorcycle Engineering in Cleveland, Ohio. The actual work was performed by one of John's Mechanics, Steve Tuttle. Steve, a M.M.I. trained mechanic from the Cleveland area, has been wrenching on Harley-Davidsons for five years.

This sequence starts as the cases are bolted together. Steve stresses the importance of inspecting the cases to be sure there is no old sealer and no nicks on the sealing surface. Steve uses 3M, number 800 sealer. He puts a bead all the way around, on the flywheel side. (he has already pre-lubed the right side bearings with oil) then he installs the other case half. The first studs to be installed are the centering studs, the 2 bottom and the center one. Steve knocks them in with a brass hammer before installing the other case bolts. The case bolts are torqued down following the sequence in the service manual. Next Steve installs the the cylinder studs to the height given in the service manual.

Note, this is a mock up case. Normally the torn threads on the cam cover boss that you see in the photos would have been repaired before the assembly started.

Fit the pistons, install the heads

The next step is to fit the pistons. If you buy pistons and cylinders as an assembly this step may already have been done for you.

Steve describes the process as follows:

1. Measure the piston according to the recommendations of the piston manufacturer.

2. Transfer that measurement, plus recommended clearance, to the dial-bore indicator.

3. Now use the Sunnen powered hone to hone the cylinder (with torque plates installed on the cylinders)

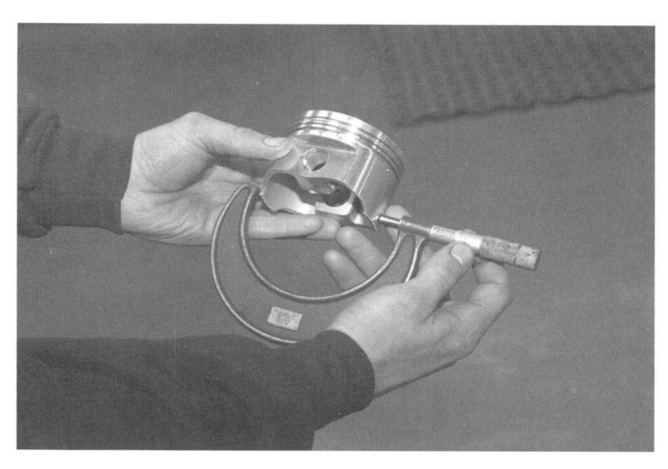

This is the start of the work at John Bryant's M-C Eng. Steve Tuttle begins the top-end assembly by measuring the piston per the manufacturer's instructions.

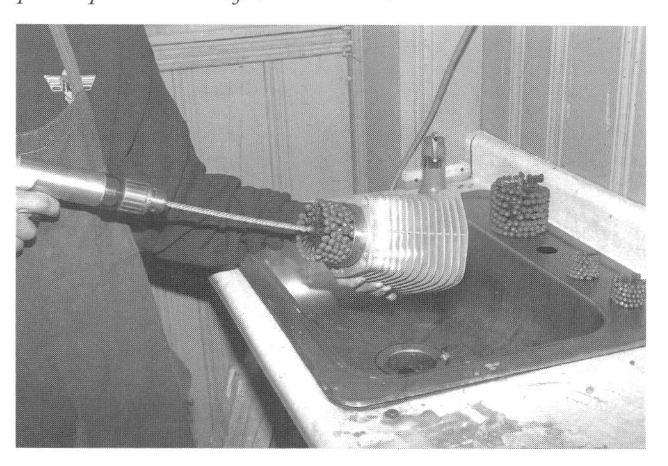

After re-sizing the cylinder, Steve uses a ball hone to create the finished pattern. The cylinder should be left with a cross-hatch pattern of 45 to 60 degrees.

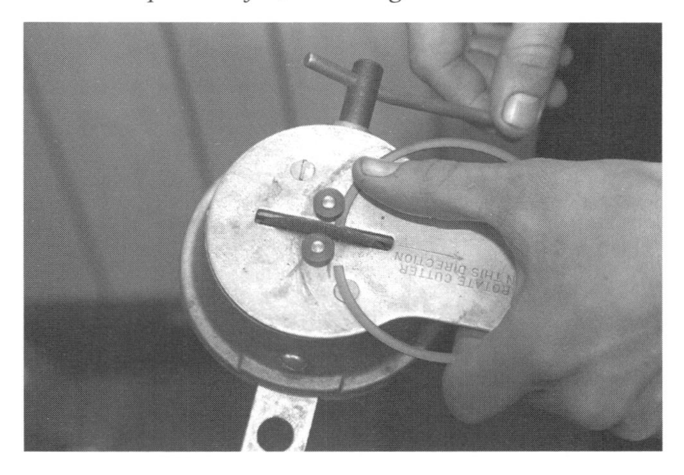

This special little grinder is designed to remove material from the end of the piston rings, though you can do the same thing (carefully) with a file.

Steve pushed the rings part way down into the cylinder bore, makes sure the rings are evenly positioned in the bore and then uses a feeler gauge to check the end gap.

to the correct size.

4. After the cylinder is the correct size, use a ball hone to create the finished 45 to 60 degree cross-hatch pattern in the cylinder. This pattern is important as it helps the cylinder hold an oil film.

5. Sand any rough edges, left from the manufacturing process, off the piston. With 320 to 400 grit sand paper sand the piston skirt in a cross-hatch pattern to obtain a pattern similar to that left on the cylinder walls by the ball hone. Be sure to thoroughly clean the pistons before installation.

6. Check the end gap of the rings according to the manufacturer's specifications.

Piston rings must be installed carefully (a special tool is made for this) because they are brittle and easy to break. They also have a top and a bottom.

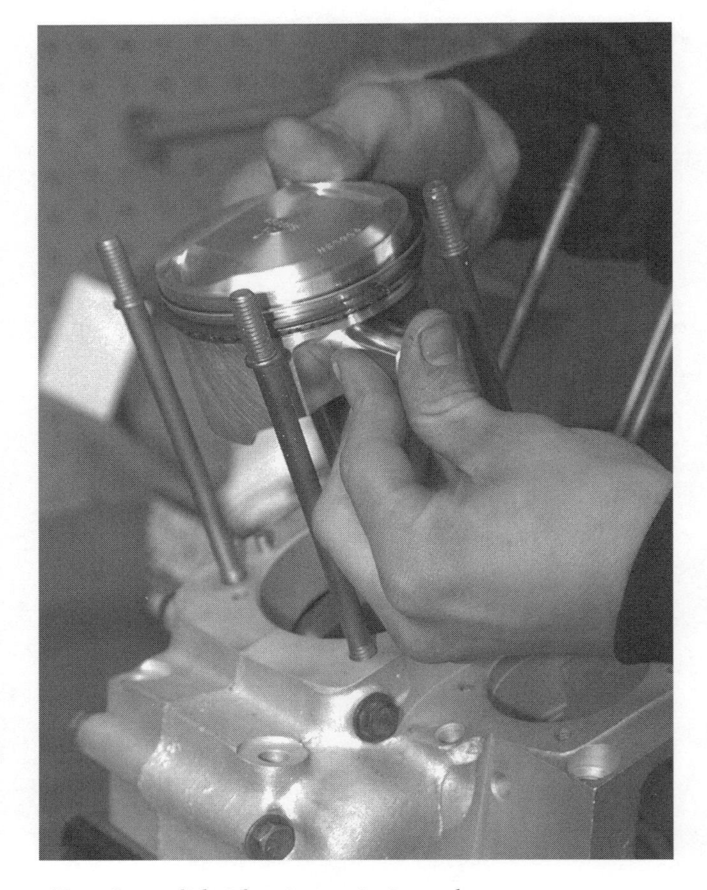

Here Steve slides the piston pin into place.

If the rings are too tight use a special little grinding tool to take some material from the end of the ring (see photo).

7. Now install the rings on the pistons per the manufacturer's instructions. Don't leave the ring gaps on the thrust face of the piston and be sure to stager the gaps so they don't "line up."

8. Install pistons, right one to the right cylinder, and make sure they are facing the right way.

9. Put in the circlip for the piston pins with rags filing the opening in the cases below the piston so that when you drop the clip (part of Murphy's law) it doesn't disappear down inside the cases.

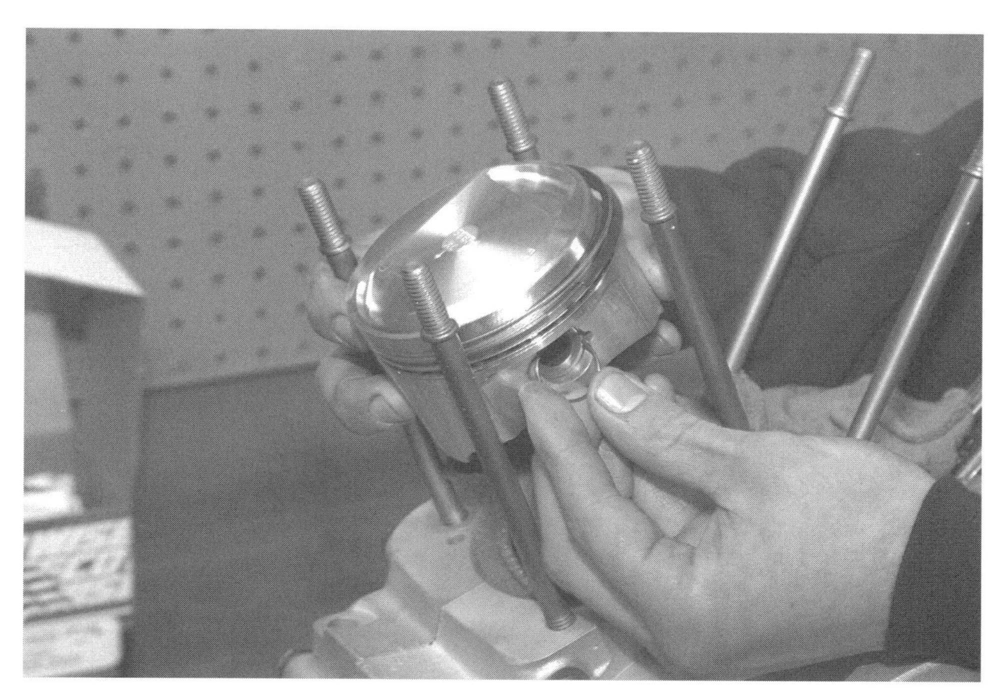

Next comes the circlips that hold the piston pin in place.

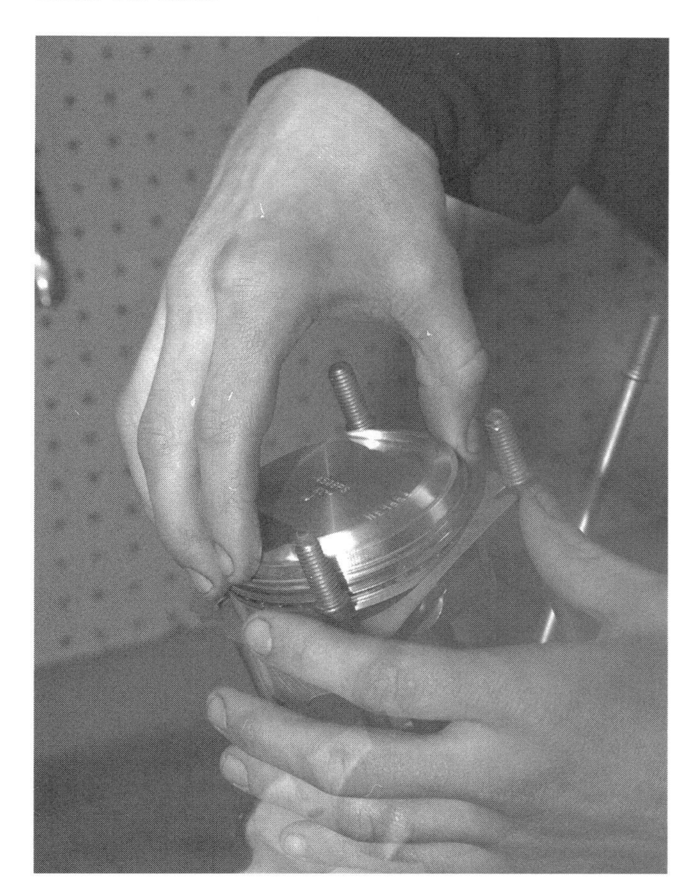

The base gasket is now slid down into place (with no sealer used).

After oiling the pistons rings and ring compressor Steve snaps the compressor over the piston rings.

O-rings are place over the dowels before the head gasket is set in place. Steve did not use any sealer on this head gasket. Note: if you look closely you can see the cross-hatch pattern left on the cylinder walls by the hone.

10. Put the base gaskets in place dry (unless the gasket manufacturer instructs you to do otherwise).

11. Now oil the compressor, piston and cylinder.

12. Compress the rings and carefully slide the cylinder down over the piston.

13. Do the same for the other cylinder.

14. Put O-rings on the cylinder locating dowels, then put on the head gasket, dry.

15. Set the heads on carefully to be sure they set down on the locating dowels.

16. Oil the threads and the bottom of the head bolts, so they don't give a false reading on the torque wrench, and torque them according to the specifications provided by the

With the compressor locked Steve carefully slides the cylinder down over the piston rings.

Steve sets the head on, making sure it drops down over the dowels.

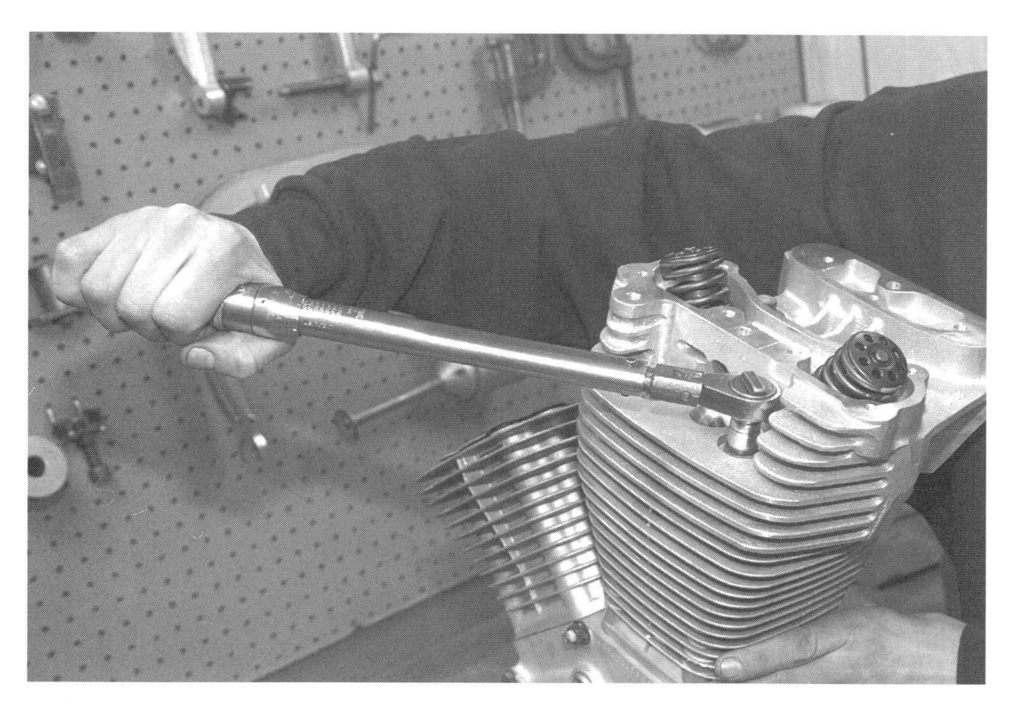

The head bolts must be tightened in a specific sequence according to instructions in the service manual.

engine manufacturer.

17. This torqueing is normally done in a three step sequence (see the service manual) to a specific pattern.

18. Now you can go ahead and install the valve train, carb and all the rest.

Note, more typically you would install the gearcase first, before the heads go on. And that way you can prime the oil system.

Steve adds the following comments for novice engine builders: Cleanliness *is* next to godliness. It really helps to be organized and neat. You should always check and double check manufacturers tolerances and specifications as you work on the motor (and don't fudge any of those). Buy good forged pistons. The actual piston clearance will vary according to the piston manufacturer.

When you fire the engine the first few times let it run only 30 to 45 seconds. Be careful how much heat it builds during those first running periods. Make the first few rides short and vary the rpm. Don't exceed 50 MPH for the first 50 to 150 miles, with varying rpm. After 100 to 150 miles then move up to highway speeds, but take it easy until 500 miles. At 500 miles, change the oil and filter and then ride faster, but don't grab a lot of throttle until 2000 miles. If all you do is a top end rebuild, you only have to take it easy for the first 500 miles.

How long a motor lasts and how well it works are both determined largely by the care that is taken in choosing and assembling the parts. Remember to follow Steve's advice to be clean, neat, organized and thorough.

Sources

Andrews Products (cams)
5212 Shapland Av
Rosemont, IL 60018
312 992 4014
FAX: 1 312 992 4017

Advanced Racing Technology
(Twin-screw blowers)
Rt 360 PO Box 247
Burgess, VA 22432

Axtell
1424 SE Maury
Des Moines, Iowa 50317
1 515 243 2518
FAX 243 0244

Best Coat
Powder Coating
1557 101st Ave. N.E.
Blaine MN 55449
612 785 7086

Brilhante Co Inc (Blower kits
for V-Twins)
3283 Motor Ave.
Los Angeles, CA 90034
310 838 5901
1 800 HOT BIKE
(Brilhante is soon to become
All American Motorcycle Co.
620 No. Lake Av, Pasadena,
CA)

Camden Superchargers
401-M East Braker Ln
Austin TX 78753
512 339 4772

Crane Cams
530 Fentress Blvd
Daytona Beach, FL 32114
904 252 1151
FAX: 904 947 5106

Custom Chrome
1 Jacqueline Court
Morgan Hill, CA 95037
408 - 778-0500

Delkron
2430 Manning St
Sacremento CA 95815
916 921 9703

Drag Specialties
9839 W 69th St
Eden Prairie, MN 55344
612 942-7890

Edelbrock QwikSilver
Carburetor Division
13465 Nomwaket Rd, Unit A
Apple Valley CA 92308
Phone: 619 247 1714
FAX: 619 247 2931,

E. T. Performance
3805 Mariana Way
Santa Barbara, CA 93105
805 563 2386

Fageol Superchargers
1255 Hayden Lane
El Cajon, CA 92021
619 447 1092
FAX 619 593 7294

Feuling R&D
2521 Palma Dr
Ventura CA 93003
805 650 2598

First Choice Turbo Center
1558 W. Henrietta Rd
East Avon, NY 14414
716 226 2929e

Deters Custom Finishing Inc
1455 91st Av NE
Blaine, MN 55449
612 784 6005

Hyperformance
5152A NE 12th Avenue
Pleasant Hill, IA 50317
515 266 6381
FAX 515 263 8008

John Bryant's Motorcycle
Engineering
16146 St. Clair
Cleveland, OH 44110
216 531 5100

Johnson Performance
Engineering
1097 Foxen Canyon Rd
Santa Maria CA 93454
805 922 3569

Keck Engineering
18887 Taylor Parkway
N. Ridgeville OH 44039
216 327 2715
FAX: 216 327 4459

Kokesh MC parts
8302 NE Hwy 65
Spring Lake Park, MN 55432
612 786 9050

Mid-USA
4937 Fyler
St. Louis MO 63139
314 351 3733
800 527 0501
FAX 314 351 6990

Merch Performance
RR2 Red Deer
Alberta, CA T4N 5E2
403 346 1221
FAX 346 3003

Mr. Turbo
(Turbo installation and kits)
4014 Hopper Rd.
Houston TX 77093
713 442 7113
FAX 713 442 4472

Nempco
7 Perry Dr.
PO Box 9137
Foxboro MA 02035
1 800 343 9687

Nitrous Oxide Systems
5930 Lakeshore Dr.
Cypress, CA 90630
714 821 0580

Performance Engineering
4327 Loraine Av
Cleveland OH 44113
216 961 5151

RB Racing
(turbo installations etc.)
1625 134th St
Gardena CA 90249
310 515 5720

Rivera Engineering
12532 Lambert Rd
Whittier CA 90606
310 907 2600
FAX: 310 907 2606

Sudco International
Distributor for Mikuni
3014 Tanager Av
Commerce CA 90040
213 728 5407

S&S Cycle
Box 215
Viola, WI 54664
608 627-1497
FAX 608 627-1488

STD Development
PO Box 3583
Chatsworth CA 91313-3583
818 998 8226
FAX: 818 998 0210

Truett & Osborn
3345 E 31st So.
Wichita, KS 67216
316 682 4781

WhiTek
PO Box 337
Arroyo Grande CA 93421-
0337
805 481 7710